General Quantum Variational Calculus

Quantum calculus is the modern name for the investigation of calculus without limits. Quantum calculus, or q-calculus, began with F.H. Jackson in the early twentieth century, but this kind of calculus had already been worked out by renowned mathematicians Euler and Jacobi.

Lately, quantum calculus has aroused a great amount of interest due to the high demand of mathematics that model quantum computing. The q-calculus appeared as a connection between mathematics and physics. It has a lot of applications in different mathematical areas such as number theory, combinatorics, orthogonal polynomials, basic hypergeometric functions and other quantum theory sciences, mechanics, and the theory of relativity. Recently, the concept of general quantum difference operators that generalize quantum calculus has been defined.

General Quantum Variational Calculus is specially designed for those who wish to understand this important mathematical concept, as the text encompasses recent developments of general quantum variational calculus. The material is presented in a highly readable, mathematically solid format. Many practical problems are illustrated, displaying a wide variety of solution techniques.

This book is addressed to a wide audience of specialists such as mathematicians, physicists, engineers, and biologists. It can be used as a textbook at the graduate level and as a reference for several disciplines.

Svetlin G. Georgiev is a mathematician who has worked in various areas of mathematics. He currently focuses on harmonic analysis, functional analysis, partial differential equations, ordinary differential equations, Clifford and quaternion analysis, integral equations, and dynamic calculus on time scales.

Khaled Zennir earned his PhD in mathematics in 2013 from Sidi Bel Abbès University, Algeria. In 2015, he received his highest diploma in Habilitation in mathematics from Constantine University, Algeria. He is currently an assistant professor at Qassim University in the Kingdom of Saudi Arabia. His research interests lie in the subjects of nonlinear hyperbolic partial differential equations: global existence, blowup, and longtime behavior.

Advances in Applied Mathematics
Series Editor: Daniel Zwillinger

Delay Ordinary and Partial Differential Equations
Andrei D. Polyanin, Vsevolod G. Sorkin, and Alexi I. Zhurov

Clean Numerical Simulation
Shijun Liao

Multiplicative Partial Differential Equations
Svetlin Georgiev and Khaled Zennir

Engineering Statistics
A Matrix-Vector Approach with MATLAB®
Lester W. Schmerr Jr.

General Quantum Numerical Analysis
Svetlin Georgiev and Khaled Zennir

An Introduction to Partial Differential Equations with MATLAB®
Matthew P. Coleman and Vladislav Bukshtynov

Handbook of Exact Solutions to Mathematical Equations
Andrei D. Polyanin

Introducing Game Theory and its Applications, Second Edition
Elliott Mendelson and Dan Zwillinger

Modeling Operations Research and Business Analytics
William P. Fox and Robert E. Burks

Decision Analysis through Modeling and Game Theory
William P. Fox

Advances in High-Order Predictive Modelling
Dan Gabriel Cacuci

Introduction to Financial Mathematics, Second Edition
Kevin J. Hastings

General Quantum Variational Calculus
Svetlin G. Georgiev and Khaled Zennir

https://www.routledge.com/Advances-in-Applied-Mathematics/book-series/CRCADVAPPMTH?pd=published,forthcoming&pg=1&pp=12&so=pub&view=list

General Quantum Variational Calculus

Svetlin G. Georgiev and Khaled Zennir

CRC Press is an imprint of the
Taylor & Francis Group, an **informa** business
A CHAPMAN & HALL BOOK

First edition published 2025
by CRC Press
2385 Executive Center Drive, Suite 320, Boca Raton, FL 33431

and by CRC Press
4 Park Square, Milton Park, Abingdon, Oxon, OX14 4RN

CRC Press is an imprint of Taylor & Francis Group, LLC

Library of Congress Cataloging-in-Publication Data
Names: Georgiev, Svetlin, author. | Zennir, Khaled, author.
Title: General quantum variational calculus / Svetlin G. Georgiev and Khaled Zennir.
Description: First edition. | Boca Raton : CRC Press, 2025. | Series: Advances in applied mathematics | Includes bibliographical references and index.
Identifiers: LCCN 2024029994 | ISBN 9781032899732 (hbk) | ISBN 9781032900698 (pbk) | ISBN 9781003546023 (ebk)
Subjects: LCSH: Calculus of variations. | Quantum theory--Mathematics.
Classification: LCC QA316 .G46 2025 | DDC 515/.64--dc23/eng/20240904
LC record available at https://lccn.loc.gov/2024029994

ISBN: 978-1-032-89973-2 (hbk)
ISBN: 978-1-032-90069-8 (pbk)
ISBN: 978-1-003-54602-3 (ebk)

DOI: 10.1201/9781003546023

Typeset in Nimbus font
by KnowledgeWorks Global Ltd.

Publisher's note: This book has been prepared from camera-ready copy provided by the authors.

Contents

Preface vii

1 **Elements of the Multimensional General Quantum Calculus** 1
1.1 The Multidimensional General Quantum Calculus 1
1.2 Line Integrals 26
1.3 The Green Formula 32
1.4 Advanced Practical Problems 34

2 **β-Differential Systems** 39
2.1 Structure of β-Differential Systems 39
2.2 β-Matrix Exponential Function 62
2.3 The β-Liouville Theorem 64
2.4 Constant Coefficients 68
2.5 Nonlinear Systems 84
2.6 Advanced Practical Problems 89

3 **Functionals** 91
3.1 Definition for Functionals 91
3.2 Self-Adjoint Second Order Matrix Equations 92
3.3 The Jacobi Condition 105
3.4 Sturmian Theory 114

4 **Linear Hamiltonian Dynamic Systems** 117
4.1 Linear Symplectic Dynamic Systems 117
4.2 Hamiltonian Systems 122
4.3 Conjoined Bases 125
4.4 Riccati Equations 136
4.5 The Picone Identity 142
4.6 "Big" Linear Hamiltonian Systems 159
4.7 Positivity of Quadratic Functionals 173

5 The First Variation 181
5.1 The Dubois-Reymond Lemma 181
5.2 The Variational Problem 188
5.3 The Euler-Lagrange Equation 196
5.4 The Legendre Condition 204
5.5 The Jacobi Condition 208
5.6 Advanced Practical Problems 211

6 Higher Order Calculus of Variations 213
6.1 Statement of the Variational Problem 213
6.2 The Euler Equation 216
6.3 Advanced Practical Problems 223

7 Double Integral Calculus of Variations 225
7.1 Statement of the Variational Problem 225
7.2 First and Second Variation 226
7.3 The Euler Condition 231
7.4 Advanced Practical Problems 236

8 The Noether Second Theorem 239
8.1 Invariance under Transformations 239
8.2 The Noether Second Theorem without Transformations of Time 243
8.3 The Noether Second Theorem with Transformations of Time 245
8.4 The Noether Second Theorem-Double Delta Integral Case 249

Bibliography 255

Index 257

Preface

Quantum calculus is the modern name for the investigation of calculus without limits. The quantum calculus, or q-calculus, began with FH Jackson in the early twentieth century, but this kind of calculus had already been worked out by Euler and Jacobi. Recently, it arose interest due to high demand for mathematics that models quantum computing. The q-calculus appeared as a connection between mathematics and physics. It has a lot of applications in different mathematical areas such as number theory, combinatorics, orthogonal polynomials, basic hypergeometric functions and other sciences quantum theory, mechanics, and the theory of relativity. The book by Kac and Cheung [3] covers many of the fundamental aspects of quantum calculus. Recently, the concept of general quantum difference operators that generalize the quantum calculus was defined.

This book encompasses recent developments of general quantum variational calculus. It is intended for use in the field of general quantum variational calculus and general quantum differential equations. It is also suitable for graduate courses in the above fields. The book contains eight chapters that are pedagogically organized. This book is specially designed for those who wish to understand general quantum variational calculus.

Chapter 1 gives the definitions for general quantum partial derivative and general quantum multidimensional integral and some of their properties are deducted. Line integrals and the Green formula are also introduced. Chapter 2 introduces general quantum differential systems. It is considered the case of constant coefficients. Chapter 3 deals with functionals and self-adjoint second order matrix equations. We also formulate and prove the Jacobi condition. It introduces Sturmian theory. Chapter 4 is concerned with linear Hamiltonian general quantum differential systems and deducts some of the basic properties of the symplectic general quantum differential systems and Hamiltonian general quantum differential systems. It also introduces Riccati equations and proves the Picone identity. Some criterions are given for positive definiteness of quadratic functionals. Chapter 5 is devoted to the first and second variation. It formulates and proves an analogue of Dubois-Reymond lemma in the general quantum case. It deducts the Euler-Lagrange equation, the Legendre condition and the Jacobi condition. Chapter 6 deals with higer order calculus of variations.

The double integral calculus of variations is considered in Chapter 7. Chapter 8 deals with the Noether second theorem. It is considered the double delta integral case of the Noether second theorem.

The aim of this book is to present a clear and well-organized treatment of the concept behind the development of mathematics and solution techniques. The text material of this book is presented in a highly readable, mathematically solid format. Many practical problems are illustrated displaying a wide variety of solution techniques.

The author welcomes any suggestions for the improvement of the text.

Paris, February 2024 *Svetlin G. Georgiev and Khaled Zennir*

Chapter 1
Elements of the Multimensional General Quantum Calculus

1.1 The Multidimensional General Quantum Calculus

Let $n \in \mathbb{N}$ be fixed. For each $i \in \{1,2,\dots,n\}$, let $I_i \subset \mathbb{R}$. Denote

$$\Lambda^n = I_1 \times I_2 \times \cdots \times I_n.$$

Definition 1.1 Let β_i, $i \in \{1,2,\dots,n\}$, be general quantum operators of first or second kind in I_i with fixed point τ_i, i.e., $\beta_i(\tau_i) = \tau_i$, $i \in \{1,\dots,n\}$. The operator $\beta : \Lambda^n \to \mathbb{R}^n$ defined by

$$\beta(t) = (\beta_1(t), \beta_2(t), \dots, \beta_n(t))$$

is said to be general quantum operator in Λ^n.

Example 1.1 Let $n = 4$ and $\Lambda^4 = [0,1] \times [0,3] \times [-1,10] \times [-5,5]$. Let also,

$$\beta_1(t) = \frac{1}{2}t, \quad t \in [0,1],$$

$$\beta_2(t) = \frac{1}{3}t + 2, \quad t \in [0,3],$$

$$\beta_3(t) = 2t - 1, \quad t \in [-1,10],$$

$$\beta_4(t) = 4t + 5, \quad t \in [-5,5].$$

Then

$$\beta(t_1,t_2,t_3,t_4) = (\beta_1(t_1), \beta_2(t_2), \beta_3(t_3), \beta_4(t_4)), \quad (t_1,t_2,t_3,t_4) \in \Lambda^4,$$

is a general quantum operator on Λ^4.

Definition 1.2 For $x = (x_1, x_2, \ldots, x_n) \in \mathbb{R}^n$ and $y = (y_1, y_2, \ldots, y_n) \in \mathbb{R}^n$, we write

$$x \geq y$$

whenever

$$x_i \geq y_i \quad \text{for all} \quad i = 1, 2, \ldots, n.$$

In a similar way, we understand $x > y$ and $x < y$ and $x \leq y$.

Definition 1.3 Let $f : \Lambda^n \to \mathbb{R}$. We introduce the following notations:

$$f^{\beta}(t) = f(\beta_1(t_1), \beta_2(t_2), \ldots, \beta_n(t_n)),$$

$$f_i^{\beta_i}(t) = f(t_1, \ldots, t_{i-1}, \beta_i(t_i), t_{i+1}, \ldots, t_n),$$

$$f_{i_1 i_2 \ldots i_l}^{\beta_{i_1} \beta_{i_2} \ldots \beta_{i_l}}(t) = f(\ldots, \beta_{i_1}(t_{i_1}), \ldots, \beta_{i_2}(t_{i_2}), \ldots, \beta_{i_l}(t_{i_l}), \ldots),$$

where $1 \leq i_1 < i_2 < \ldots < i_l \leq n$, $i_m \in \mathbb{N}$, $m \in \{1, 2, \ldots, l\}$, $l \in \mathbb{N}$.

Example 1.2 Let $n = 2$ and

$$I_1 = [-1, 1],$$

$$I_2 = [0, 5]$$

and

$$\beta_1(t_1) = \frac{1}{3} t_1, \quad t_1 \in I_1,$$

$$\beta_2(t_2) = \frac{1}{4} t_2 + 3, \quad t_2 \in I_2,$$

are general quantum operators in I_1 and I_2, respectively. Then

$$\beta(t) = (\beta_1(t_1), \beta_2(t_2)), \quad t = (t_1, t_2) \in \Lambda^2 = I_1 \times \mathbb{I}_2$$

is a general quantum operator in Λ^2. Consider

$$f(t) = t_1^2 - 2t_1 t_2, \quad t = (t_1, t_2) \in \Lambda^2.$$

Then

$$\begin{aligned}
f^{\beta}(t) &= (\beta_1(t_1))^2 - 2\beta_1(t_1)\beta_2(t_2)\\
&= \left(\frac{1}{3}t_1\right)^2 - 2\left(\frac{1}{3}t_1\right)\left(\frac{1}{4}t_2+3\right)\\
&= \frac{1}{9}t_1^2 - \frac{2}{3}t_1\left(\frac{1}{4}t_2+3\right)\\
&= \frac{1}{9}t_1^2 - \frac{1}{6}t_1t_2 - 2t_1, \quad t=(t_1,t_2)\in\Lambda^2,
\end{aligned}$$

and

$$\begin{aligned}
f^{\beta_1}(t) &= (\beta_1(t_1))^2 - 2\beta_1(t_1)t_2\\
&= \left(\frac{1}{3}t_1\right)^2 - \frac{2}{3}t_1t_2\\
&= \frac{1}{9}t_1^2 - \frac{2}{3}t_1t_2, \quad t=(t_1,t_2)\in\Lambda^2,
\end{aligned}$$

and

$$\begin{aligned}
f^{\beta_2}(t) &= t_1^2 - 2t_1\beta_2(t_2)\\
&= t_1^2 - 2t_1\left(\frac{1}{4}t_2+3\right)\\
&= t_1^2 - \frac{1}{2}t_1t_2 - 6t_1, \quad t=(t_1,t_2)\in\Lambda^2.
\end{aligned}$$

Example 1.3 Let $\Lambda^2 = (2\mathbb{Z}\cup\{-1\})\times\left(2^{\mathbb{N}}\cup\{0\}\right)$. Here,

$$I_1 = 2\mathbb{Z},$$

$$I_2 = 2^{\mathbb{N}}\cup\{0\}.$$

Let also,

$$\beta_1(t_1) = t_1+2, \quad t_1\in I_1, \quad t_1\neq -1,$$

$$\beta_1(-1) = -1,$$

$$\beta_2(t_2) = 2t_2, \quad t_2\in I_2.$$

Take

$$f(t_1,t_2) = t_1^2 + t_2, \quad (t_1,t_2) \in \Lambda^2.$$

Hence,

$$\begin{aligned}
f^{\beta}(t) &= f(\beta_1(t_1), \beta_2(t_2)) \\
&= (\beta_1(t_1))^2 + \beta_2(t_2) \\
&= (t_1+2)^2 + 2t_2 \\
&= t_1^2 + 4t_1 + 2t_2 + 4, \\
f_1^{\beta_1}(t) &= f(\beta_1(t_1), t_2) \\
&= (\beta_1(t_1))^2 + t_2 \\
&= (t_1+2)^2 + t_2 \\
&= t_1^2 + 4t_1 + t_2 + 4, \\
f_2^{\beta_2}(t) &= f(t_1, \beta_2(t_2)) \\
&= t_1^2 + \beta_2(t_2) \\
&= t_1^2 + 2t_2, \quad t = (t_1,t_2) \in \Lambda^2, \quad t_1 \neq -1.
\end{aligned}$$

Exercise 1.1 Let $\Lambda^3 = (\mathbb{N} \cup \{-1\}) \times (\mathbb{N} \cup \{-1\}) \times (3\mathbb{N} \cup \{-1\})$ and $f : \Lambda^3 \to \mathbb{R}$ be defined as

$$f(t) = t_1 t_2 t_3, \quad t = (t_1,t_2,t_3) \in \Lambda^3.$$

Let also,

$$\begin{aligned}
\beta_1(t_1) &= t_1 + 1, \quad t_1 \in \mathbb{N}, \quad t_1 \neq -1, \\
\beta_1(-1) &= -1, \\
\beta_2(t_2) &= t_2 + 1, \quad t_2 \in \mathbb{N}, \quad t_2 \neq -1, \\
\beta_2(-1) &= -1, \\
\beta_3(t_3) &= t_3 + 3, \quad t_3 \in 3\mathbb{N}, \quad t_3 \neq -1, \\
\beta_3(-1) &= -1.
\end{aligned}$$

Find

1. $f^{\beta}(t), t \in \Lambda^3$.
2. $f_1^{\beta_1}(t), t \in \Lambda^3$.
3. $f_2^{\beta_2}(t), t \in \Lambda^3$.
4. $f_3^{\beta_3}(t), t \in \Lambda^3$.
5. $f_{12}^{\beta_1\beta_2}(t), t \in \Lambda^3$.
6. $f_{13}^{\beta_1\beta_3}(t), t \in \Lambda^3$.
7. $f_{23}^{\beta_2\beta_3}(t), t \in \Lambda^3$.
8. $f^{\beta}(t) + f_1^{\beta_1}(t), t \in \Lambda^3$.

Definition 1.4 Assume that $f : \Lambda^n \to \mathbb{R}$ is a function and let $t \in \Lambda_i$. We define

$$\begin{aligned}\frac{\partial f(t_1,t_2,\ldots,t_n)}{\partial \beta_i} &= \frac{\partial f(t)}{\partial \beta_i}\\ &= \frac{\partial f}{\partial \beta_i}(t)\\ &= f_{\beta_i}(t)\\ &= f_{t_i}^{D\beta_i}\end{aligned}$$

to be the number, provided it exists, with the property that for any $\varepsilon_i > 0$, there exists a neighborhood

$$U_i = (t_i - \delta_i, t_i + \delta_i) \cap I_i,$$

for some $\delta_i > 0$, such that

$$\Big| f(t_1,\ldots,t_{i-1},\beta_i(t_i),t_{i+1},\ldots,t_n) - f(t_1,\ldots,t_{i-1},s_i,t_{i+1},\ldots,t_n)$$

$$- f_{\beta_i}(t)(\beta_i(t_i) - s_i)\Big| \leq \varepsilon_i |\beta_i(t_i) - s_i| \quad \text{for all} \quad s_i \in U_i. \quad (1.1)$$

We call $f_{\beta_i}(t)$ the general partial derivative (or partial derivative) of f with respect to t_i at t. We say that f is partial differentiable (or general partial differentiable) with respect to t_i in Λ_i if $f_{\beta_i}(t)$ exists for all $t \in \Lambda_i$. The function $f_{\beta_i} : \Lambda_i \to \mathbb{R}$ is said to be the general partial derivative (or partial derivative) with respect to t_i of f in Λ_i.

Note that the general partial derivative is well defined.

Example 1.4 Let $f(t) = t_1t_2t_3$, $t = (t_1,t_2,t_3) \in \Lambda^3$. We will prove that

$$f_{\beta_1}(t) = t_2t_3.$$

Indeed, for every $\varepsilon_1 > 0$, there exists $\delta_1 > 0$ such that for every $s_1 \in (t_1 - \delta_1, t_1 + \delta_1)$, $s_1 \in I_1$, we have

$$
\begin{aligned}
&|f(\beta_1(t_1), t_2, t_3) - f(s_1, t_2, t_3) - t_2 t_3 (\beta_1(t_1) - s_1)| \\
&= |\beta_1(t_1) t_2 t_3 - s_1 t_2 t_3 - t_2 t_3 (\beta_1(t_1) - s_1)| \\
&= 0 \\
&\leq \varepsilon_1 |\beta_1(t_1) - s_1|.
\end{aligned}
$$

Remark 1.1 For $t \in \Lambda^n$ and $s_i \in I_i$, we write

$$t_{s_i} = (t_1, \ldots, t_{i-1}, s_i, t_{i+1}, \ldots, t_n), \quad i \in \{1, 2, \ldots, n\}.$$

Then we can rewrite (1.1) in the form

$$|f^{\beta_i}(t) - f(t_{s_i}) - f_{\beta_i}(t)(\beta_i(t_i) - s_i)| \leq \varepsilon_i |\beta_i(t_i) - s_i|. \tag{1.2}$$

The general partial derivative has the following properties:

1. Let $f : \Lambda^n \to \mathbb{R}$ be a function and $t \in \Lambda_i$. If f is general partial differentiable with respect to t_i at t, then
$$\lim_{s_i \to t_i} f(t_{s_i}) = f(t). \tag{1.3}$$
2. Let $f : \Lambda^n \to \mathbb{R}$, $t \in \Lambda_i$, and
$$\lim_{s_i \to t_i} f(t_{s_i}) = f(t). \tag{1.4}$$
If $t_i \neq s_i$, then f is general partial differentiable with respect to t_i at t and
$$f_{\beta_i}(t) = \frac{f^{\beta_i}(t) - f(t)}{\beta_i(t_i) - t_i}. \tag{1.5}$$

Example 1.5 Let $\Lambda^2 = (\mathbb{N} \cup \{0\}) \times (2^{\mathbb{N}} \cup \{0\})$ and define $f : \Lambda^2 \to \mathbb{R}$ by

$$f(t) = t_1^2 t_2 - 2t_1, \quad t = (t_1, t_2) \in \Lambda^2,$$

and

$$
\begin{aligned}
\beta_1(t_1) &= t_1 + 1, \quad t_1 \in I_1, \quad t_1 \neq 0, \\
\beta_1(0) &= 0, \\
\beta_2(t_2) &= 2t_2, \quad t_2 \in I_2.
\end{aligned}
$$

Therefore,

$$
\begin{aligned}
f_{t_1}^{D_{\beta_1}}(t) &= \frac{f_1^{\beta_1}(t)-f(t)}{\beta_1(t_1)-t_1} \\
&= \frac{(\beta_1(t_1))^2 t_2 - 2\beta_1(t_1) - t_1^2 t_2 + 2t_1}{t_1+1-t_1} \\
&= (t_1+1)^2 t_2 - 2(t_1+1) - t_1^2 t_2 + 2t_1 \\
&= t_1^2 t_2 + 2t_1 t_2 + t_2 - 2t_1 - 2 - t_1^2 t_2 + 2t_1 \\
&= 2t_1 t_2 + t_2 - 2, \quad (t_1,t_2) \in \Lambda_1^{\kappa_1 2}, \quad t_1 \neq 0,
\end{aligned}
$$

$$
\begin{aligned}
f_{t_2}^{D_{\beta_2}}(t) &= \frac{f_2^{\beta_2}(t)-f(t)}{\beta_2(t_2)-t_2} \\
&= \frac{t_1^2\beta_2(t_2) - 2t_1 - (t_1^2 t_2 - 2t_1)}{2t_2 - t_2} \\
&= \frac{2t_1^2 t_2 - 2t_1 - t_1^2 t_2 + 2t_1}{t_2} \\
&= t_1^2, \quad (t_1,t_2) \in \Lambda_2^{\kappa_2 2}, \quad t_1 \neq 0.
\end{aligned}
$$

Theorem 1.1 *Let $t \in \Lambda_i^{\kappa_i n}$ and $t_i = \beta_i(t_i)$. Then f is partial delta differentiable with respect to t_i at t if and only if the limit*

$$\lim_{s_i \to t_i} \frac{f(t)-f(t_{s_i})}{t_i - s_i}$$

exists as a finite number. In this case,

$$f_{t_i}^{D_{\beta_i}}(t) = \lim_{s_i \to t_i} \frac{f(t)-f(t_{s_i})}{t_i - s_i}. \tag{1.6}$$

Example 1.6 Let

$$\Lambda^2 = \left\{\frac{1}{2}, 1, 2\right\} \times (\mathbb{N} \cup \{0\})$$

and define $f : \Lambda^2 \to \mathbb{R}$ by

$$f(t) = t_1^2 t_2, \quad t = (t_1,t_2) \in \Lambda^2,$$

and

$$\beta_1\left(\frac{1}{2}\right) = \frac{1}{2},$$

$$\beta_1(1) = 2,$$

$$\beta_1(2) = 3,$$

$$\beta_2(t_2) = t_2 + 1, \quad t_2 \in I_2, \quad t_2 \neq 0,$$

$$\beta_2(0) = 0.$$

We will find $f_{t_1}^{D_{\beta_1}}\left(\frac{1}{2}, t_2\right)$, $t_2 \in I_2$, $t_2 \neq 0$. We have,

$$\begin{aligned}
\lim_{s_1 \to \frac{1}{2}} \frac{f\left(\frac{1}{2}, t_2\right) - f(s_1, t_2)}{\frac{1}{2} - s_1} &= \lim_{s_1 \to \frac{1}{2}} \frac{\frac{1}{4}t_2 - s_1^2 t_2}{\frac{1}{2} - s_1} \\
&= \lim_{s_1 \to \frac{1}{2}} \frac{\left(\frac{1}{2} - s_1\right)\left(\frac{1}{2} + s_1\right) t_2}{\frac{1}{2} - s_1} \\
&= \lim_{s_1 \to \frac{1}{2}} \left(\frac{1}{2} + s_1\right) t_2 \\
&= t_2.
\end{aligned}$$

Consequently,

$$f_{t_1}^{D_{\beta_1}}\left(\frac{1}{2}, t_2\right) = t_2.$$

Exercise 1.2 Let $\Lambda^2 = (\mathbb{N} \cup \{-1\}) \times (4\mathbb{N} \cup \{1\}$ and define $f : \Lambda^2 \to \mathbb{R}$ by

$$f(t) = t_1^3 + 3t_1^2 + t_1 t_2, \quad t = (t_1, t_2) \in \Lambda^2,$$

and

$$\beta_1(t_1) = t_1 + 1, \quad t_1 \in I, \quad t_1 \neq -1,$$

$$\beta_1(-1) = -1,$$

$$\beta_2(t_2) = t_2 + 4, \quad t_2 \in I_2, \quad t_2 \neq 1,$$

$$\beta_2(1) = 1.$$

Find $f_{t_1}^{D_{\beta_1}}(1, t_2)$.

Theorem 1.2 *Let $t \in \Lambda^n$. Suppose $f : \Lambda^n \to \mathbb{R}$ is a function that is partial differentiable with respect to t_i at t. If $\alpha \in \mathbb{R}$, then αf is partial differentiable with respect to t_i at t and*

$$(\alpha f)_{t_i}^{D_{\beta_i}}(t) = \alpha f_{t_i}^{D_{\beta_i}}(t).$$

Theorem 1.3 *Let $t \in \Lambda^n$. Assume $f, g : \Lambda^n \to \mathbb{R}$ are partial differentiable with respect to t_i at t. Then $f+g$ is partial differentiable with respect to t_i at t and*

$$(f+g)_{t_i}^{D_{\beta_i}}(t) = f_{t_i}^{D_{\beta_i}}(t) + g_{t_i}^{D_{\beta_i}}(t).$$

Theorem 1.4 *Let $t \in \Lambda^n$. Assume $f, g : \Lambda^n \to \mathbb{R}$ are partial differentiable with respect to t_i at t. Then fg is partial differentiable with respect to t_i at t and*

$$(fg)_{t_i}^{D_{\beta_i}}(t) = f_{t_i}^{D_{\beta_i}}(t)g(t) + f_i^{\beta_i}(t)g_{t_i}^{D_{\beta_i}}(t) = f(t)g_{t_i}^{D_{\beta_i}}(t) + f_{t_i}^{D_{\beta_i}}(t)g_i^{\beta_i}(t).$$

Example 1.7 Let $\Lambda^2 = (\mathbb{N} \cup \{-1\}) \times \left(\mathbb{N}_0^2 \cup \left\{\frac{1}{4}\right\}\right)$ and define $h : \Lambda^2 \to \mathbb{R}$ by

$$h(t) = (t_1^2 + 2t_1)(t_1^3 + t_2), \quad t \in \Lambda^2,$$

and

$$\beta_1(t_1) = t_1 + 1, \quad t_1 \in I_1, \quad t_1 \neq -1,$$

$$\beta_1(-1) = -1,$$

$$\beta_2(t_2) = (1 + \sqrt{t_2})^2, \quad t_2 \in I_2, \quad t_2 \neq \frac{1}{4},$$

$$\beta_2\left(\frac{1}{4}\right) = \frac{1}{4}.$$

We will find $h_{t_1}^{D_{\beta_1}}(t)$, $t \in \Lambda^2$. Let

$$f(t) = t_1^2 + 2t_1,$$

$$g(t) = t_1^3 + t_2, \quad t \in \Lambda^2.$$

Hence,

$$h(t) = f(t)g(t), \quad t \in \Lambda^2.$$

Then, for $t \in \Lambda^2$, $t_1 \neq -1$, $t_2 \neq \frac{1}{4}$, we have

$$\begin{aligned}
h_{t_1}^{D_{\beta_1}}(t) &= f_{t_1}^{D_{\beta_1}}(t)g(t) + f_1^{\beta_1}(t)g_{t_1}^{D_{\beta_1}}(t) \\
&= (\beta_1(t_1) + t_1 + 2)g(t) + ((\beta_1(t_1))^2 + 2\beta_1(t_1))((\beta_1(t_1))^2 + t_1\beta_1(t_1) + t_1^2) \\
&= (t_1 + 1 + t_1 + 2)(t_1^3 + t_2) + ((t_1+1)^2 + 2t_1 + 2)((t_1+1)^2 + t_1(t_1+1) + t_1^2) \\
&= (2t_1 + 3)(t_1^3 + t_2) + (t_1^2 + 4t_1 + 3)(3t_1^2 + 3t_1 + 1) \\
&= 2t_1^4 + 2t_1t_2 + 3t_1^3 + 3t_2 + 3t_1^4 + 3t_1^3 + t_1^2 + 12t_1^3 + 12t_1^2 + 4t_1 + 9t_1^2 + 9t_1 + 3 \\
&= 5t_1^4 + 18t_1^3 + 22t_1^2 + 13t_1 + 2t_1t_2 + 3t_2 + 3, \quad t \in \Lambda^2.
\end{aligned}$$

Theorem 1.5 *Let $f, g : \Lambda^n \to \mathbb{R}$ be partial differentiable with respect to t_i at $t \in \Lambda^n$. Assume $g_i^{\beta_i}(t)g(t) \neq 0$. Then $\frac{f}{g}$ is partial differentiable with respect to t_i at t and*

$$\left(\frac{f}{g}\right)_{t_i}^{D_{\beta_i}}(t) = \frac{f_{t_i}^{D_{\beta_i}}(t)g(t) - f(t)g_{t_i}^{D_{\beta_i}}(t)}{g_i^{\beta_i}(t)g(t)}.$$

Exercise 1.3 Let $\Lambda^2 = (2\mathbb{N} \cup \{-1\}) \times (\mathbb{N}_0 \cup \{-1\})$ and define $h : \Lambda^2 \to \mathbb{R}$ by

$$h(t) = \frac{t_1 + t_2}{t_1 - t_2 + 1}, \quad t \in \Lambda^2.$$

Let also,

$$\beta_1(t_1) = t_1 + 2, \quad t_1 \in I_1, \quad t_1 \neq -1,$$

$$\beta_1(-1) = -1,$$

$$\beta_2(t_2) = t_2 + 1, \quad t_2 \in I_2, \quad t_2 \neq -1,$$

$$\beta_2(-1) = -1.$$

Find $h_{t_1}^{D_{\beta_1}}(t)$ for $t \in \Lambda_1^{\kappa_1 2}$, $t_1 \neq -1$, $t_2 \neq -1$, $(t_1 - t_2 + 2)(t_1 - t_2 + 1) \neq 0$, and $h_{t_2}D_{\beta_2}(t)$ for $t \in \Lambda_2^{\kappa_2 2}$, $(t_1 - t_2)(t_1 - t_2 + 1) \neq 0$.

Definition 1.5 For a function $f : \Lambda^n \to \mathbb{R}$, we shall talk about the second order partial derivative with respect to t_i and t_j, $i, j \in \{1, 2, \ldots, n\}$, $f_{t_i t_j}^{D_{\beta_i} D_{\beta_j}}$, provided $f_{t_i}^{D_{\beta_i}}$ is partial differentiable with respect to t_j on $\Lambda^n = (\Lambda^n)^n$ with partial derivative

$$f_{t_i t_j}^{D_{\beta_i} D_{\beta_j}} = \left(f_{t_i}^{D_{\beta_i}}\right)_{t_j}^{D_{\beta_j}} : \Lambda^n \to \mathbb{R}.$$

For $i = j$, we will write

$$f_{t_i t_i}^{D_{\beta_i} D_{\beta_i}} = f_{t_i^2}^{{D_{\beta_i}}^2}.$$

Similarly, we define higher order partial derivatives

$$f_{t_i t_j \ldots t_l}^{D_{\beta_i} D_{\beta_j} \ldots D_{\beta_l}} : \Lambda^n \to \mathbb{R}.$$

For $t \in \Lambda^n$, we define

$$\beta^2(t) = \beta(\beta(t)) = (\beta_1(\beta_1(t_1)), \beta_2(\beta_2(t_2)), \ldots, \beta_n(\beta_n(t_n)))$$

and

$$(\beta^{-1})^2(t) = \beta^{-1}(\beta^{-1}(t)) = (\beta_1^{-1}(\beta_1^{-1}(t_1)), \beta_2^{-1}(\beta_2^{-1}(t_2)), \ldots, \beta_n^{-1}(\beta_n^{-1}(t_n))),$$

and $\beta^m(t)$ and $(\beta^{-1})^m(t)$ for $m \in \mathbb{N}$ are defined accordingly. Finally, we put

$$(\beta^{-1})^0(t) = \beta^0(t) = t, \quad f_{t_i}^{{D_{\beta_i}}^0} = f.$$

Example 1.8 Let $\Lambda^3 = (\mathbb{N} \cup \{0\}) \times \mathbb{N}_0 \times \mathbb{Z}$. Here

$$I_1 = \mathbb{N}, \quad I_2 = \mathbb{N}_0, \quad I_3 = \mathbb{Z}$$

and

$$\beta_1(t_1) = t_1 + 1, \quad t_1 \in I_1, \quad t_1 \neq 0,$$

$$\beta_1(0) = 0,$$

$$\beta_2(t_2) = t_2 + 1, \quad t_2 \in I_2, \quad t_2 \neq 0,$$

$$\beta_2(0) = 0,$$

$$\beta_3(t_3) = t_3 + 1, \quad t_3 \in I_3, \quad t_3 \neq 0,$$

$$\beta_3(0) = 0.$$

Hence,

$$\beta^2(t) = (\beta_1(\beta_1(t_1)), \beta_2(\beta_2(t_2)), \beta_3(\beta_3(t_3)))$$

$$= (\beta_1(t_1)+1, \beta_2(t_2)+1, \beta_3(t_3)+1)$$

$$= (t_1+1+1, t_2+1+1, t_3+1+1)$$

$$= (t_1+2, t_2+2, t_3+2), \quad t \in \Lambda^3, \quad t_1, t_2, t_3 \neq 0.$$

Exercise 1.4 Let $\Lambda^2 = (3^{\mathbb{N}} \cup \{1\}) \times \mathbb{N}$. Find $\beta^2(t)$ and $(\beta^{-1})^2(t)$, $t \in \Lambda^2$, where

$$\beta_1(t_1) = 3t_1, \quad t_1 \in I_1, \quad t_1 \neq 1,$$

$$\beta_1(1) = 1,$$

$$\beta_2(t_2) = t_2 + 1, \quad t_2 \in I_2, \quad t_2 \neq 0,$$

$$\beta_2(0) = 0.$$

Theorem 1.6 (Leibniz Formula) *Let $S_{ik}^{(m)}$ be the set consisting of all possible strings of length m, containing exactly k times β_i and $m-k$ times D_{β_i}. If $f^{\alpha}_{t_i^{m-k}}$ exists for any $\alpha \in S_{ik}^{(m)}$ and $g_{t_i^k}^{D_{\beta_i}{}^k}$ exists for any $k \in \{0, 1, \ldots, m\}$, then*

$$(fg)_{t_i^m}^{D_{\beta_i}{}^m} = \sum_{k=0}^{m} \left(\sum_{\alpha \in S_{ik}^{(m)}} f^{\alpha}_{t_i^{m-k}} \right) g_{t_i^k}^{D_{\beta_i}{}^k} \tag{1.7}$$

holds for any $m \in \mathbb{N}$.

Remark 1.2 We note that (1.7) holds for $m = 0$ with the convention that

$$\sum_{\alpha \in \emptyset} f^{\alpha}_{t_i^k} = f.$$

Definition 1.6 We put

$$\Lambda^n_{\kappa} = I_{1\kappa} \times I_{2\kappa} \times \ldots \times I_{n\kappa},$$
$$\Lambda^n_{i\kappa_i} = I_1 \times \ldots \times I_{i-1} \times I_{i\kappa} \times I_{i+1} \times \ldots \times I_n, \quad i = 1, 2, \ldots, n,$$
$$\Lambda^n_{i_1 i_2 \ldots i_l \kappa_{i_1} \kappa_{i_2} \ldots \kappa_{i_l}} = \ldots \times I_{i_1\kappa} \times \ldots \times I_{i_2\kappa} \times \ldots \times I_{i_l\kappa} \times \ldots,$$

where $1 \leq i_1 < i_2 < \ldots < i_l \leq n$, $i_m \in \mathbb{N}$, $m = 1, 2, \ldots, l$.

Now we will introduce the conception for general quantum multiple integration. Assume that I_i, $i \in \{1,\ldots,n\}$, are closed subsets of $\mathbb{R}$. Suppose $a_i < b_i$ are points in I_i and $[a_i,b_i)$ is the half-closed bounded interval in I_i, $i \in \{1,\ldots,n\}$. Let us introduce a "rectangle" in $\Lambda^n = I_1 \times I_2 \times \ldots \times I_n$ by

$$R = [a_1,b_1) \times [a_2,b_2) \times \ldots [a_n,b_n)$$

$$= \{(t_1,t_2,\ldots,t_n) : t_i \in [a_i,b_i),\ i = 1,2,\ldots,n\}.$$

Let

$$a_i = t_i^0 < t_i^1 < \ldots < t_i^{k_i} = b_i.$$

Definition 1.7 We call the collection of intervals

$$P_i = \left\{[t_i^{j_i-1}, t_i^{j_i}) : j_i = 1,\ldots,k_i\right\}, \quad i = 1,2,\ldots,n,$$

a β_i-partition of $[a_i,b_i)$ and denote the set of all β_i-partitions of $[a_i,b_i)$ by $P_i([a_i,b_i))$.

Definition 1.8 Let

$$R_{j_1 j_2 \cdots j_n} = [t_1^{j_1-1}, t_1^{j_1}) \times [t_2^{j_2-1}, t_2^{j_2}) \times \ldots \times [t_n^{j_n-1}, t_n^{j_n})$$
$$1 \le j_i \le k_i, \quad i = 1,2,\ldots,n. \tag{1.8}$$

We call the collection

$$P = \left\{R_{j_1 j_2 \ldots j_n} : 1 \le j_i \le k_i,\ i = 1,2,\ldots,n\right\} \tag{1.9}$$

a β-partition of R, generated by the β_i-partitions P_i of $[a_i,b_i)$, and we write

$$P = P_1 \times P_2 \times \ldots \times P_n.$$

The set of all β-partitions of R is denoted by $\mathscr{P}(R)$. Moreover, for a bounded function $f : R \to \mathbb{R}$, we set

$$M = \sup\{f(t_1,t_2,\ldots,t_n) : (t_1,t_2,\ldots,t_n) \in R\},$$

$$m = \inf\{f(t_1,t_2,\ldots,t_n) : (t_1,t_2,\ldots,t_n) \in R\},$$

$$M_{j_1 j_2 \cdots j_n} = \sup\{f(t_1,t_2,\ldots,t_n) : (t_1,t_2,\ldots,t_n) \in R_{j_1 j_2 \cdots j_n}\},$$

$$m_{j_1 j_2 \cdots j_n} = \inf\{f(t_1,t_2,\ldots,t_n) : (t_1,t_2,\ldots,t_n) \in R_{j_1 j_2 \cdots j_n}\}.$$

Definition 1.9 The upper Darboux β-sum $U(f,P)$ and the lower Darboux β-sum $L(f,P)$ with respect to P are defined by

$$U(f,P) = \sum_{j_1=1}^{k_1}\sum_{j_2=1}^{k_2}\cdots\sum_{j_n=1}^{k_n} M_{j_1j_2\cdots j_n}(t_1^{j_1}-t_1^{j_1-1})(t_2^{j_2}-t_2^{j_2-1})\dots(t_n^{j_n}-t_n^{j_n-1})$$

and

$$L(f,P) = \sum_{j_1=1}^{k_1}\sum_{j_2=1}^{k_2}\cdots\sum_{j_n=1}^{k_n} m_{j_1j_2\cdots j_n}(t_1^{j_1}-t_1^{j_1-1})(t_2^{j_2}-t_2^{j_2-1})\dots(t_n^{j_n}-t_n^{j_n-1}).$$

Remark 1.3 We note that

$$\begin{aligned} U(f,P) &\le M\sum_{j_1=1}^{k_1}\sum_{j_2=1}^{k_2}\cdots\sum_{j_n=1}^{k_n}(t_1^{j_1}-t_1^{j_1-1})(t_2^{j_2}-t_2^{j_2-1})\dots(t_n^{j_n}-t_n^{j_n-1}) \\ &\le M(b_1-a_1)(b_2-a_2)\dots(b_n-a_n) \end{aligned}$$

and

$$\begin{aligned} L(f,P) &\ge m\sum_{j_1=1}^{k_1}\sum_{j_2=1}^{k_2}\cdots\sum_{j_n=1}^{k_n}(t_1^{j_1}-t_1^{j_1-1})(t_2^{j_2}-t_2^{j_2-1})\dots(t_n^{j_n}-t_n^{j_n-1}) \\ &\ge M(b_1-a_1)(b_2-a_2)\dots(b_n-a_n), \end{aligned}$$

i.e.,

$$\begin{aligned} m(b_1-a_1)(b_2-a_2)\dots(b_n-a_n) \le & L(f,P) \\ \le & U(f,P) \\ \le & M(b_1-a_1)(b_2-a_2)\dots(b_n-a_n). \end{aligned} \tag{1.10}$$

Definition 1.10 The upper Darboux β-integral $U(f)$ of f over R and the lower Darboux β-integral $L(f)$ of f over R are defined by

$$U(f) = \inf\{U(f,P) : P \in \mathscr{P}(R)\} \quad \text{and} \quad L(f) = \sup\{L(f,P) : P \in \mathscr{P}(R)\}.$$

From (1.10), it follows that $U(f)$ and $L(f)$ are finite real numbers.

Definition 1.11 We say that f is β-integrable over R provided $L(f) = U(f)$. In this case, we write

$$\int_R f(t_1,t_2,\dots,t_n)D_{\beta_1}t_1\dots D_{\beta_n}t_n$$

for this common value. We call this integral the Darboux β-integral.

Remark 1.4 For a given rectangle

$$V = [c_1, d_1) \times [c_2, d_2) \times \ldots \times [c_n, d_n) \subset \Lambda^n,$$

the "area" of V, i.e., $(d_1 - c_1)(d_2 - c_2)\ldots(d_n - c_n)$, is denoted by $m(V)$.

Definition 1.12 Let $P, Q \in \mathscr{P}(R)$ and

$$P = P_1 \times P_2 \times \ldots \times P_n, \quad Q = Q_1 \times Q_2 \times \ldots \times Q_n,$$

where $P_i, Q_i \in \mathscr{P}([a_i, b_i))$. We say that Q is a refinement of P provided Q_i is a refinement of P_i for all $i \in \{1, 2, \ldots, n\}$.

Theorem 1.7 *Let f be a bounded function on R. If P and Q are β-partitions of R and Q is a refinement of P, then*

$$L(f, P) \le L(f, Q) \le U(f, Q) \le U(f, P),$$

i.e., refining of a partition increases the lower sum and decreases the upper sum.

Definition 1.13 Suppose

$$P = P_1 \times P_2 \times \ldots \times P_n \quad \text{and} \quad Q = Q_1 \times Q_2 \times \ldots \times Q_n,$$

where $P_i, Q_i \in \mathscr{P}([a_i, b_i))$, $i \in \{1, 2, \ldots, n\}$, are two β-partitions of

$$R = [a_1, b_1) \times [a_2, b_2) \times \ldots \times [a_n, b_n).$$

If P_i is generated by a set

$$\{t_i^0, t_i^1, \ldots, t_i^{k_i}\}, \quad \text{where} \quad a_i = t_i^0 < t_i^1 < \ldots < t_i^{k_i} = b_i,$$

and Q_i is generated by a set

$$\{\tau_i^0, \tau_i^2, \ldots, \tau_i^{p_i}\}, \quad \text{where} \quad a_i = \tau_i^0 < \tau_i^1 < \ldots < \tau_i^{p_i} = b_i,$$

then, by

$$P + Q = (P_1 + Q_1) \times (P_2 + Q_2) \times \ldots \times (P_n + Q_n),$$

we denote the β-partition of R generated by

$$P_i + Q_i = \{t_i^0, t_i^1, \ldots, t_i^{k_i}\} \cup \{\tau_i^0, \tau_i^1, \ldots, \tau_i^{p_i}\}, \quad i = 1, 2, \ldots, n.$$

Remark 1.5 Obviously, $P + Q$ is a refinement of both P and Q.

Theorem 1.8 *If f is a bounded function on R, and if P and Q are any two β-partitions of R, then*

$$L(f, P) \le U(f, Q),$$

i.e., every lower sum is less than or equal to every upper sum.

Theorem 1.9 *If f is a bounded function on R, then $L(f) \leq U(f)$.*

Theorem 1.10 *If $L(f,P) = U(f,P)$ for some $P \in \mathscr{P}(R)$, then the function f is β-integrable over R and*

$$\int_R f(t_1,t_2,\ldots,t_n)D_{\beta_1}t_1 \ldots D_{\beta_n}t_n = L(f,P) = U(f,P).$$

Theorem 1.11 *A bounded function f on R is β-integrable if and only if for each $\varepsilon > 0$, there exists $P \in \mathscr{P}(R)$ such that*

$$U(f,P) - L(f,P) < \varepsilon. \tag{1.11}$$

Definition 1.14 We denote by $P_\delta([a,b))$ the set of all $P_i \in \mathscr{P}([a,b))$ that possess the property

$$t_i^{j_l} - t_i^{j_{l-1}} < \delta$$

or

$$t_i^{j_l} - t_i^{j_{l-1}} \geq \delta and t_i^{j_l} = \beta_i(t_i^{j_{l-1}}).$$

Further, by $\mathscr{P}_\delta(R)$, we denote the set of all $P \in \mathscr{P}(R)$ such that

$$P = P_1 \times P_2 \times \ldots \times P_n, \quad \text{where} \quad P_i \in \mathscr{P}_\delta([a_i,b_i)), \quad i = 1,2,\ldots,n.$$

Theorem 1.12 *Let $P^0 \in \mathscr{P}(R)$ be given by*

$$P^0 = P_1^0 \times P_2^0 \times \ldots \times P_n^0$$

in which $P_i^0 \in \mathscr{P}([a_i,b_i))$, $i \in \{1,2,\ldots,n\}$, is generated by a set

$$A_i^0 = \{t_{i_0}^0, t_{i_1}^0, \ldots, t_{i_{n_i}}^0\} \subset [a_i,b_i], \quad where \quad a_i = t_{i_0}^0 < t_{i_1}^0 < \ldots < t_{i_{n_i}}^0 = b_i.$$

Then, for each $P \in \mathscr{P}_\delta(R)$, we have

$$L(f,P^0+P) - L(f,P) \leq (M-m)D^{n-1}(n_1+n_2+\ldots+n_n-n)\delta$$

and

$$U(f,P) - U(f,P+P^0) \leq (M-m)D^{n-1}(n_1+n_2+\ldots+n_n-n)\delta,$$

where $D = \max\limits_{i\in\{1,2,\ldots,n\}} \{b_i - a_i\}$, and M and m are defined as above.

Theorem 1.13 *A bounded function f on R is β-integrable if and only if for each $\varepsilon > 0$, there exists $\delta > 0$ such that*

$$P \in \mathscr{P}_\delta(R) \quad implies \quad U(f,P) - L(f,P) < \varepsilon. \tag{1.12}$$

Theorem 1.14 *For every bounded function f on R, the Darboux β-sums $L(f,P)$ and $U(f,P)$ evaluated for $P \in \mathscr{P}_\delta(R)$ have limits as $\delta \to 0$, uniformly with respect to P, and*

$$\lim_{\delta \to 0} L(f,P) = L(f) \quad \text{and} \quad \lim_{\delta \to 0} U(f,P) = U(f).$$

Definition 1.15 Let f be a bounded function on R and $P \in \mathscr{P}(R)$. In each "rectangle" $R_{j_1 j_2 \dots j_n}$, $1 \le j_i \le k_i$, $i = 1,2,\dots,n$, choose a point $\xi_{j_1 j_2 \dots j_n}$ and form the sum

$$S = \sum_{i=1}^{n} \sum_{j_i=1}^{k_i} f(\xi_{j_1 j_2 \dots j_n})(t_1^{j_i} - t_1^{j_i-1}) \dots (t_n^{j_n} - t_n^{j_n-1}). \tag{1.13}$$

We call S a Riemann β-sum of f corresponding to $P \in \mathscr{P}(R)$. We say that f is Riemann β-integrable over R if there exists a number I such that, for each $\varepsilon > 0$, there exists $\delta > 0$ such that

$$|S - I| < \varepsilon$$

for every Riemann β-sum S of f corresponding to any $P \in \mathscr{P}_\delta(R)$, independent of the choice of the point $\xi_{j_1 j_2 \dots j_n} \in R_{j_1 j_2 \dots j_n}$ for $1 \le j_i \le k_i$, $i = 1,2,\dots,n$. The number I is called the Riemann β-integral of f over R. We write

$$I = \lim_{\delta \to 0} S.$$

Theorem 1.15 *The Riemann β-integral is well defined.*

Remark 1.6 Note that in the Riemann definition of the integral, we need not assume the boundedness of f in advance. However, it follows that the Riemann integrability of a function f over R implies its boundedness on R.

Theorem 1.16 *A bounded function on R is Riemann β-integrable if and only if it is Darboux β-integrable, in which case the values of the integrals are equal.*

Remark 1.7 In the definition of

$$\int_R f(t_1,t_2,\dots,t_n) D_{\beta_1} t_1 \dots D_{\beta_n} t_n$$

with $R = [a_1,b_1) \times \dots \times [a_n,b_n)$, we assumed that $a_i < b_i$, $i \in \{1,\dots,n\}$. We extend the definition to the case $a_i = b_i$ for some $i \in \{1,2,\dots,n\}$ by setting

$$\int_R f(t_1,t_2,\dots,t_n) D_{\beta_1} t_1 \dots D_{\beta_n} t_n = 0 \tag{1.14}$$

if $a_i = b_i$ for some $i \in \{1,\dots,n\}$.

Theorem 1.17 *Let*

$$a = (a_1,\dots,a_n) \in \Lambda^n \quad \text{and} \quad b = (b_1,\dots,b_n) \in \Lambda^n$$

with $a_i \leq b_i$ for all $i \in \{1, \dots, n\}$. Every constant function

$$f(t_1, t_2, \dots, t_n) = A \quad for \quad (t_1, t_2, \dots, t_n) \in R = [a_1, b_1) \times \dots \times [a_n, b_n)$$

is β-integrable over R and

$$\int_R f(t_1, t_2, \dots, t_n) D_{\beta_1} t_1 \dots D_{\beta_n} t_n = A \prod_{i=1}^{n} (b_i - a_i). \tag{1.15}$$

Theorem 1.18 *Let $t^0 = (t_1^0, \dots, t_n^0) \in \Lambda^n$. Every function $f : \Lambda^n \to \mathbb{R}$ is β-integrable over*

$$R = R(t^0) = [t_1^0, \beta_1(t_1^0)) \times \dots \times [t_n^0, \beta_n(t_n^0)),$$

and

$$\int_R f(t_1, t_2, \dots, t_n) D_{\beta_1} t_1 \dots D_{\beta_n} t_n = \prod_{i=1}^{n} (\beta_i(t_i^0) - t_i^0) f(t_1^0, \dots, t_n^0). \tag{1.16}$$

Theorem 1.19 *Let*

$$a = (a_1, \dots, a_n) \in \Lambda^n \quad and \quad b = (b_1, \dots, b_n) \in \Lambda^n$$

with $a_i \leq b_i$ for all $i \in \{1, 2, \dots, n\}$. If $I_i = \mathbb{R}$ for every $i \in \{1, 2, \dots, n\}$, then every bounded function f on $R = [a_1, b_1) \times [a_2, b_2) \times \dots \times [a_n, b_n)$ is β-integrable if and only if f is Riemann integrable on R in the classical sense, and in this case

$$\int_R f(t_1, t_2, \dots, t_n) D_{\beta_1} t_1 \dots D_{\beta_n} t_n = \int_R f(t_1, t_2, \dots, t_n) \mathrm{d}t_1 \mathrm{d}t_2 \dots \mathrm{d}t_n,$$

where the integral on the right-hand side is the ordinary Riemann integral.

Theorem 1.20 *Let*

$$a = (a_1, \dots, a_n) \in \Lambda^n \quad and \quad b = (b_1, \dots, b_n) \in \Lambda^n$$

with $a_i \leq b_i$ for all $i \in \{1, 2, \dots, n\}$. If $I_i = \mathbb{Z}$ for all $i \in \{1, 2, \dots, n\}$, then every function defined on $R = [a_1, b_1) \times [a_2, b_2) \times \dots \times [a_n, b_n)$ is β-integrable over R, and

$$\int_R f(t_1, t_2, \dots, t_n) D_{\beta_1} t_1 \dots D_{\beta_n} t_n$$

$$= \begin{cases} 0 & \text{if } a_i = b_i \text{ for some } i \in \{1, 2, \dots, n\} \\ \sum\limits_{r_1 = a_1}^{b_1 - 1} \sum\limits_{r_2 = a_2}^{b_2 - 1} \dots \sum\limits_{r_n = a_n}^{b_n - 1} f(r_1, r_2, \dots, r_n) & \text{otherwise.} \end{cases} \tag{1.17}$$

Note that Λ^n is a complete metric space with the metric d defined by

$$d(t,s) = \sqrt{\sum_{i=1}^{n}(t_i - s_i)^2} \quad \text{for} \quad t = (t_1, t_2, \dots, t_n), \quad s = (s_1, s_2, \dots, s_n) \in \Lambda^n,$$

and also with the equivalent metric

$$d(t,s) = \max_{i \in \{1,2,\dots,n\}} \{|t_i - s_i|\}.$$

Definition 1.16 A function $f : \Lambda^n \to \mathbb{R}$ is said to be continuous at $t \in \Lambda^n$ if for every $\varepsilon > 0$, there exists $\delta > 0$ such that

$$|f(t) - f(s)| < \varepsilon$$

for all points $s \in \Lambda^n$ satisfying $d(t,s) < \delta$.

Remark 1.8 If t is an isolated point of Λ^n, then every function $f : \Lambda^n \to \mathbb{R}$ is continuous at t. In particular, if $I_i = \mathbb{Z}$ for all $i \in \{1, 2, \dots, n\}$, then every function $f : \Lambda^n \to \mathbb{R}$ is continuous at each point of Λ^n.

Theorem 1.21 *Every continuous function on $K = [a_1, b_1] \times [a_2, b_2] \times \dots \times [a_n, b_n]$ is β-integrable over $R = [a_1, b_1) \times [a_2, b_2) \times \dots \times [a_n, b_n)$.*

Definition 1.17 We say that a function $\phi : [\alpha, \gamma] \to \mathbb{R}$ satisfies the Lipschitz condition if there exists a constant $B > 0$, a so-called Lipschitz constant, such that

$$|\phi(u) - \phi(v)| \leq B|u - v| \quad \text{for all} \quad u, v \in [\alpha, \gamma].$$

Example 1.9 Let $\phi : [0, 1] \to \mathbb{R}$ be defined by $\phi(x) = x^2 + 1$, $x \in [0, 1]$. Then, for $x, y \in [0, 1]$, we have

$$\begin{aligned} |\phi(x) - \phi(y)| &= |x^2 + 1 - y^2 - 1| \\ &= |x^2 - y^2| \\ &= |x - y||x + y| \\ &\leq |x - y|(|x| + |y|) \\ &\leq 2|x - y|, \end{aligned}$$

i.e., ϕ satisfies the Lipschitz condition with Lipschitz constant $B = 2$.

Theorem 1.22 *Let f be bounded and β-integrable over*

$$R = [a_1, b_1) \times [a_2, b_2) \times \dots \times [a_n, b_n)$$

and let M and m be its supremum and infimum over R, respectively. If $\phi : [m,M] \to \mathbb{R}$ is a function satisfying the Lipschitz condition, then the composite function $h = \phi \circ f$ is β-integrable over R.

Theorem 1.23 *Let f be a bounded function that is β-integrable over*

$$R = [a_1, b_1) \times [a_2, b_2) \times \ldots \times [a_n, b_n).$$

If $a_i', b_i' \in [a_i, b_i]$ with $a_i' < b_i'$ for all $i \in \{1, 2, \ldots, n\}$, then f is β-integrable over $R' = [a_1', b_1') \times [a_2', b_2') \times \ldots \times [a_n', b_n')$.

Theorem 1.24 *Let f be a bounded function that is β-integrable on*

$$R = [a_1, b_1) \times [a_2, b_2) \times \ldots \times [a_n, b_n).$$

If $\alpha \in \mathbb{R}$, then αf is β-integrable on R and

$$\int_R \alpha f(t_1, t_2, \ldots, t_n) D_{\beta_1} t_1 \ldots D_{\beta_n} t_n = \alpha \int_R f(t_1, t_2, \ldots, t_n) D_{\beta_1} t_1 \ldots D_{\beta_n} t_n. \tag{1.18}$$

Theorem 1.25 *If f and g are bounded functions that are β-integrable over*

$$R = [a_1, b_1) \times [a_2, b_2) \times \ldots \times [a_n, b_n),$$

then $f + g$ is β-integrable over R and

$$\int_R (f+g)(t_1, t_2, \ldots, t_n) D_{\beta_1} t_1 \ldots D_{\beta_n} t_n$$

$$= \int_R f(t_1, t_2, \ldots, t_n) D_{\beta_1} t_1 \ldots D_{\beta_n} t_n + \int_R g(t_1, t_2, \ldots, t_n) D_{\beta_1} t_1 \ldots D_{\beta_n} t_n. \tag{1.19}$$

Theorem 1.26 *If f and g are bounded functions that are β-integrable over R with*

$$f(t_1, t_2, \ldots, t_n) \le g(t_1, t_2, \ldots, t_n) \quad \text{for all} \quad (t_1, t_2, \ldots, t_n) \in R,$$

then

$$\int_R f(t_1, t_2, \ldots, t_n) D_{\beta_1} t_1 \ldots D_{\beta_n} t_n \le \int_R g(t_1, t_2, \ldots, t_n) D_{\beta_1} t_1 \ldots D_{\beta_n} t_n.$$

Theorem 1.27 *If f is a bounded function that is β-integrable over R, then so is $|f|$, and*

$$\left| \int_R f(t_1, t_2, \ldots, t_n) D_{\beta_1} t_1 \ldots D_{\beta_n} t_n \right| \le \int_R |f(t_1, t_2, \ldots, t_n)| D_{\beta_1} t_1 \ldots D_{\beta_n} t_n.$$

Theorem 1.28 *If f is a bounded function that is β-integrable over R, then so is f^2.*

Theorem 1.29 *If f and g are β-integrable over R, then so is fg.*

Theorem 1.30 *Let the rectangle R be the union of two disjoint rectangles R_1 and R_2. If f is a bounded function that is β-integrable on each of R_1 and R_2, then f is β-integrable over R and*

$$\int_R f(t_1,t_2,\dots,t_n)D_{\beta_1}t_1\dots D_{\beta_n}t_n = \int_{R_1} f(t_1,t_2,\dots,t_n)D_{\beta_1}t_1\dots D_{\beta_n}t_n + \int_{R_2} f(t_1,t_2,\dots,t_n)D_{\beta_1}t_1\dots D_{\beta_n}t_n.$$

Now we will extend the definition for the multiple Riemann β-integral over more general sets in Λ^n, called Jordan β-measurable sets.

Definition 1.18 Let $E \subset \Lambda^n$. A point $t = (t_1,t_2,\dots,t_n) \in \Lambda^n$ is called a boundary point of E if every open ball $B(t,r) = \{x \in \Lambda^n : d(t,x) < r\}$ of radius r with center t contains at least one point of $E \setminus \{t\}$ and at least one point of $\Lambda^n \setminus E$. The set of all boundary points of E is called the boundary of E and is denoted by ∂E.

Definition 1.19 Let $E \subset \Lambda^n$. A point $t = (t_1,t_2,\dots,t_n) \in \Lambda^n$ is called a β-boundary point of E if every rectangle of the form

$$V = [t_1,t_1') \times [t_2,t_2') \times \dots \times [t_n,t_n') \subset \Lambda^n$$

with $t_i' \in I_i$, $t_1' > t_i$, $i = 1,2,\dots,n$, contains at least one point of E and at least one point of $\Lambda^n \setminus E$. The set of all β-boundary points of E is called the β-boundary and is denoted by $\partial_\beta E$.

For $i \in \{1,2,\dots,n\}$, we set $I_i^0 = I_i \setminus \{\max I_i\}$. If I_i does not have a maximum, then $I_i^0 = I_i$, $i \in \{1,2,\dots,n\}$. Evidently, for every point $t_i \in I_i^0$, there exists an interval $[\alpha_i,\beta_i) \subset I_i$ that contains the point t_i, $i \in \{1,2,\dots,n\}$.

Let $\Lambda_0^n = I_1^0 \times I_2^0 \times \dots \times I_n^0$.

Definition 1.20 Let $E \subset \Lambda_0^n$ be a bounded set and let $\partial_\beta E$ be its boundary. Assume

$$R = [a_1,b_1) \times [a_2,b_2) \times \dots \times [a_n,b_n)$$

is a rectangle in Λ^n such that $E \cup \partial_\beta E \subset R$. Suppose $\mathscr{P}(R)$ denotes the set of all β-partitions of R of the type (1.8) and (1.9). For every $P \in \mathscr{P}(R)$, denote with $J_*(E,P)$ the sum of the areas of those subrectangles of P which are entirely contained in E, and let $J^*(E,P)$ be the sum of the areas of the subrectangles of P each of which contains at least one point of $E \cup \partial_\beta E$. The numbers

$$J_*(E) = \sup\{J_*(E,P) : P \in \mathscr{P}(R)\} \quad \text{and} \quad J^*(E) = \in f\{J^*(E,P) : P \in \mathscr{P}(R)\}$$

are called the inner and outer Jordan β-measure of E, respectively. The set E is said to be Jordan β-measurable if $J_*(E) = J^*(E)$, in which case this common value is called the Jordan β-measure of E, denoted by $J(E)$.

We note that

$$0 \leq J_*(E) \leq J^*(E)$$

for every bounded set $E \subset \Lambda_0^n$. If E has Jordan β-measure zero, then $J_*(E) = J^*(E) = 0$. Therefore, we have the following statement.

Proposition 1.1 *A bounded set $E \subset \Lambda_0^n$ has Jordan β-measure zero if and only if for every $\varepsilon > 0$, the set E can be covered by a finite collection of rectangles of type*

$$V_j = [\alpha_1^j, \beta_1^j) \times [\alpha_2^j, \beta_2^j) \times \ldots \times [\alpha_n^j, \beta_n^j) \subset \Lambda^n, \quad j = 1, 2, \ldots, m,$$

the sum of whose areas is less than ε:

$$E \subset \bigcup_{j=1}^{m} V_j \quad \textit{and} \quad \sum_{j=1}^{m} m(V_j) < \varepsilon.$$

If E is a set of Jordan β-measure zero, then so is any set $\tilde{E} \subset E$.

Proposition 1.2 *The union of a finite number of bounded subsets $E_1, E_2, \ldots, E_k \subset \Lambda_0^n$ each of which has Jordan β-measure zero is in turn a set of Jordan β-measure zero.*

The empty set is a Jordan β-measurable set, and its Jordan β-measure is zero.

Theorem 1.31 *For each point $t^0 = (t_1^0, t_2^0, \ldots, t_n^0) \in \Lambda_0^n$, the single point set $\{t^0\}$ is Jordan β-measurable, and its Jordan β-measure is given by*

$$J(\{t^0\}) = (\beta_1(t_1^0) - t_1^0)(\beta_2(t_2^0) - t_2^0) \ldots (\beta_n(t_n^0) - t_n^0). \tag{1.20}$$

Theorem 1.32 *If $E \subset \Lambda_0^n$ is a bounded set with β-boundary $\partial_\beta E$, then*

$$J^*(\partial_\beta E) = J^*(E) - J_*(E). \tag{1.21}$$

Moreover, E is Jordan β-measurable iff its β-boundary $\partial_\beta E$ has Jordan β-measure zero.

Proposition 1.3 *Let $E_1, E_2 \in \Lambda^n$. Then we have the following relations:*

1. $\partial_\beta(E_1 \cup E_2) \subset \partial_\beta E_1 \cup \partial_\beta E_2$,
2. $\partial_\beta(E_1 \cap E_2) \subset \partial_\beta E_1 \cup \partial_\beta E_2$,
3. $\partial_\beta(E_1 \setminus E_2) \subset \partial_\beta E_1 \cup \partial_\beta E_2$.

Theorem 1.33 *The union of a finite number of Jordan β-measurable sets is a Jordan β-measurable set.*

Theorem 1.34 *The intersection of a finite number of Jordan β-measurable sets is a Jordan β-measurable set.*

Theorem 1.35 *The difference of two Jordan β-measurable sets is a Jordan β-measurable set.*

Definition 1.21 Let f be defined and bounded on a bounded Jordan β-measurable set $E \subset \Lambda_0^n$. Let

$$R = [a_1, b_1) \times [a_2, b_2) \times \ldots \times [a_n, b_n) \subset \Lambda^n$$

be a rectangle containing E and put

$$K = [a_1, b_1] \times [a_2, b_2] \times \ldots \times [a_n, b_n].$$

Define f on K by

$$F(t) = \begin{cases} f(t) & \text{if} \quad t \in E \\ 0 & \text{if} \quad t \in K \setminus E. \end{cases} \tag{1.22}$$

(here $t = (t_1, t_2, \ldots, t_n)$). In this case, f is said to be Riemann β-integrable over E if F is Riemann β-integrable over R. We write

$$\int_E f(t_1, t_2, \ldots, t_n) D_{\beta_1} t_1 \ldots D_{\beta_n} t_n = \int_R F(t_1, t_2, \ldots, t_n) D_{\beta_1} t_1 \ldots D_{\beta_n} t_n.$$

Definition 1.22 Assume that f is defined and bounded on a bounded Jordan β-measurable set $E \subset \Lambda_0^n$. Let $R = [a_1, b_1) \times [a_2, b_2) \times \ldots \times [a_n, b_n) \subset \Lambda^n$ be a rectangle such that $E \subset R$. Let $P \in \mathscr{P}(R)$ be given by (1.8) and (1.9). Some of the rectangles of P will be entirely within E, some will be outside of E, and some will be partly within and partly outside E. Let $P' = \{R_1, R_2, \ldots, R_k\}$ be the collection of subrectangles in P that lie completely within the set E. The collection P' is called the inner β-partition of the set E, determined by the partition P of the rectangle R. We choose an arbitrary point $\xi^l = (\xi_1^l, \xi_2^l, \ldots, \xi_n^l)$ in the subrectangle R_l of P' for $l \in \{1, 2, \ldots, k\}$. Let $m(R_l)$ denote the area of R_l. We set

$$S = \sum_{l=1}^{k} f(\xi^l) m(R_l).$$

We call S a Riemann β-sum of f corresponding to the partition $P \in \mathscr{P}(R)$. We say that f is Riemann β-integrable over $E \subset \Lambda_0^n$ if there exists a number I such that for each $\varepsilon > 0$, there exists $\delta > 0$ such that $|S - I| < \varepsilon$ for every Riemann β-sum S of f corresponding to any inner β-partition $P' = \{R_1, R_2, \ldots, R_k\}$ of E, determined by a partition $P \in \mathscr{P}_\delta(R)$, independent of the way in which $\xi^l \in R_l$ for $1 \le l \le k$ is chosen. The number I is called the Riemann multiple β-integral of f over E. We write $I = \lim_{\delta \to 0} S$.

Proposition 1.4 *Let $E \subset \Lambda_0^n$ be a bounded set with β-boundary $\partial_\beta E$. If*

$$R = [a_1, b_1) \times [a_2, b_2) \times \ldots \times [a_n, b_n) \subset \Lambda^n$$

is a rectangle that contains $E \cup \partial_\beta E$, then

$$\lim_{\delta \to 0} J_*(E, P) = J_*(E) \quad \text{and} \quad \lim_{\delta \to 0} J^*(E, P) = J^*(E)$$

for any $P \in \mathscr{P}_\delta(R)$.

Proposition 1.5 *Let $\Gamma \subset \Lambda_0^n$ be a set of Jordan β-measure zero. Moreover, let*

$$R = [a_1, b_1) \times [a_2, b_2) \times \ldots \times [a_n, b_n)$$

be a rectangle in Λ^n that contains Γ. Then, for each $\varepsilon > 0$, there exists $\delta > 0$ such that for every partition $P \in \mathscr{P}_\delta(R)$, the sum of areas of subrectangles of P which have a common point with Γ is less than ε.

Theorem 1.36 *If $E \subset \Lambda_0^n$ is a bounded and Jordan β-measurable set and f is a bounded function on E, then Definition 1.21 and Definition 1.22 of the Riemann β-integrability of f over E are equivalent to each other.*

Theorem 1.37 *If $E \subset \Lambda_0^n$ is a bounded and Jordan β-measurable set, then the integral*

$$\int_E 1 D_{\beta_1} t_1 \ldots D_{\beta_n} t_n$$

exists. Moreover, we have

$$J(E) = \int_E 1 D_{\beta_1} t_1 \ldots D_{\beta_n} t_n. \tag{1.23}$$

Theorem 1.38 *Let $E_1, E_2 \subset \Lambda_0^n$ be bounded Jordan β-measurable sets such that $J(E_1 \cap E_2) = 0$, and let $E = E_1 \cup E_2$. If $f : E \to \mathbb{R}$ is a bounded function which is β-integrable over each of E_1 and E_2, then f is β-integrable over E and*

$$\int_E f(t_1, t_2, \ldots, t_n) D_{\beta_1} t_1 \ldots D_{\beta_n} t_n = \int_{E_1} f(t_1, t_2, \ldots, t_n) D_{\beta_1} t_1 \ldots D_{\beta_n} t_n + \int_{E_2} f(t_1, t_2, \ldots, t_n) D_{\beta_1} t_1 \ldots D_{\beta_n} t_n.$$

Exercise 1.5 Let f and g be β-integrable over E, and let $\alpha, \gamma \in \mathbb{R}$. Prove that $\alpha f + \gamma g$ is also β-integrable over E and

$$\int_E (\alpha f + \gamma g)(t_1, t_2, \ldots, t_n) D_{\beta_1} t_1 \ldots D_{\beta_n} t_n = \alpha \int_E f(t_1, t_2, \ldots, t_n) D_{\beta_1} t_1 \ldots D_{\beta_n} t_n + \gamma \int_E g(t_1, t_2, \ldots, t_n) D_{\beta_1} t_1 \ldots D_{\beta_n} t_n.$$

Exercise 1.6 Let f and g be β-integrable over E. Prove that so is their product fg.

Exercise 1.7 Let f be β-integrable over E. Prove that so is $|f|$ with

$$\left| \int_E f(t_1,t_2,\ldots,t_n) D_{\beta_1} t_1 \ldots D_{\beta_n} t_n \right| \le \int_E |f(t_1,t_2,\ldots,t_n)| D_{\beta_1} t_1 \ldots D_{\beta_n} t_n.$$

Exercise 1.8 Let f and g be β-integrable over E and

$$f(t_1,t_2,\ldots,t_n) \le g(t_1,t_2,\ldots,t_n) \quad \text{for all} \quad (t_1,t_2,\ldots,t_n) \in E.$$

Prove that

$$\int_E f(t_1,t_2,\ldots,t_n) D_{\beta_1} t_1 \ldots D_{\beta_n} t_n \le \int_E g(t_1,t_2,\ldots,t_n) D_{\beta_1} t_1 \ldots D_{\beta_n} t_n.$$

Exercise 1.9 Let f and g be β-integrable over E and let g be nonnegative (or non-positive) on E. Prove that there exists a real number

$$\Lambda \in [m,M]$$

such that

$$\int_E f(t_1,t_2,\ldots,t_n) g(t_1,t_2,\ldots,t_n) D_{\beta_1} t_1 \ldots D_{\beta_n} t_n$$

$$= \Lambda \int_E g(t_1,t_2,\ldots,t_n) D_{\beta_1} t_1 \ldots D_{\beta_n} t_n,$$

where

$$m = \inf f\{f(t_1,t_2,\ldots,t_n) : (t_1,t_2,\ldots,t_n) \in E\}$$

and

$$M = \sup\{f(t_1,t_2,\ldots,t_n) : (t_1,t_2,\ldots,t_n) \in E\}.$$

Theorem 1.39 *Let $[a_i,b_i] \subset I_i^0$, $i = 1,2,\ldots,n-1$, and let*

$$\phi : [a_1,b_1] \times [a_2,b_2] \times \ldots \times [a_{n-1},b_{n-1}] \to I_n^0$$

and

$$\psi : [a_1,b_1] \times [a_2,b_2] \times \ldots \times [a_{n-1},b_{n-1}] \to I_n^0$$

be continuous functions such that $\phi(t) < \psi(t)$ for all

$$t = (t_1,t_2,\ldots,t_{n-1}) \in [a_1,b_1] \times [a_2,b_2] \times \ldots \times [a_{n-1},b_{n-1}].$$

Let E be the bounded set in Λ^n given by

$$E = \Big\{ (t_1,t_2,\ldots,t_n) \in \Lambda^n : \ a_i \le t_i < b_i, \ i \in \{1,2,\ldots,n-1\},$$
$$\phi(t_1,t_2,\ldots,t_{n-1}) \le t_n < \psi(t_1,t_2,\ldots,t_{n-1}) \Big\}.$$

In this case, E is Jordan β-measurable, and if $f : E \to \mathbb{R}$ is β-integrable over E and if the single integral

$$\int_{\phi(t_1,t_2,\dots,t_{n-1})}^{\psi(t_1,t_2,\dots,t_{n-1})} f(t_1,t_2,\dots,t_{n-1},t_n)D_{\beta_n}t_n$$

exists for each $(t_1,t_2,\dots,t_{n-1}) \in [a_1,b_1) \times [a_2,b_2) \times \dots \times [a_{n-1},b_{n-1})$, then the iterated integral

$$\int_{a_1}^{b_1}\int_{a_2}^{b_2}\cdots\int_{a_{n-1}}^{b_{n-1}} D_{\beta_1}t_1 D_{\beta_2}t_2 \dots D_{\beta_{n-1}}t_{n-1} \int_{\phi(t_1,t_2,\dots,t_{n-1})}^{\psi(t_1,t_2,\dots,t_{n-1})} f(t_1,t_2,\dots,t_n)D_{\beta_n}t_n$$

exists. Moreover, we have

$$\int_E f(t_1,t_2,\dots,t_n)D_{\beta_1}t_1 \dots D_{\beta_n}t_n$$

$$= \int_{a_1}^{b_1}\int_{a_2}^{b_2}\cdots\int_{a_{n-1}}^{b_{n-1}} D_{\beta_1}t_1 D_{\beta_2}t_2 \dots D_{\beta_{n-1}}t_{n-1} \int_{\phi(t_1,t_2,\dots,t_{n-1})}^{\psi(t_1,t_2,\dots,t_{n-1})} f(t_1,t_2,\dots,t_n)D_{\beta_n}t_n.$$

1.2 Line Integrals

Let $I \subset \mathbb{R}$ and β be a first kind or second kind quantum operator. Let $a,b \in \mathbb{T}$ with $a < b$. Assume that $\phi_i : [a,b] \to \mathbb{R}$ is continuous, $i \in \{1,\dots,m\}$.

Definition 1.23 The system of functions

$$x_i = \phi_i(t), \quad t \in [a,b] \subset \mathbb{T}, \quad i \in \{1,\dots,m\} \tag{1.24}$$

is said to define a continuous curve Γ. The points $A(x_1,x_2,\dots,x_m)$ with the coordinates $x_1,x_2,\dots,x_m$ given by (1.24) are called the points of the curve. The set of all points of the curve will be referred to as the curve. The points

$$A_0(\phi_1(a),\dots,\phi_m(a)) \quad \text{and} \quad A_1(\phi_1(b),\dots,\phi_m(b))$$

are called the initial point and the final point of the curve, respectively. A_0 and A_1 are called the end points of the curve.

Example 1.10 The system

$$x_1 = t^2+t, \quad x_2 = t-2, \quad x_3 = t, \quad t \in [0,2],$$

defines a continuous curve. Here,

$$\phi_1(t) = t^2 + t, \quad \phi_2(t) = t - 2, \quad \phi_3(t) = t.$$

Thus,

$$\phi_1(0) = 0, \quad \phi_2(0) = -2, \quad \phi_3(0) = 0,$$

$$\phi_1(2) = 6, \quad \phi_2(2) = 0, \quad \phi_3(2) = 2.$$

Moreover, $A_0(0,-2,0)$ is the initial point, and $A_1(6,0,2)$ is the final point.

Definition 1.24 If the initial and final points coincide, then the curve is said to be closed.

Example 1.11 Consider the curve

$$x_1 = t^2 - 3t, \quad x_2 = t^3 - 4t^2 + 5t, \quad t \in [1,2].$$

Here,

$$\phi_1(t) = t^2 - 3t, \quad \phi_2(t) = t^3 - 4t^2 + 5t.$$

Thus,

$$\phi_1(1) = -2, \quad \phi_2(1) = 2,$$

$$\phi_1(2) = -2, \quad \phi_2(2) = 2.$$

Hence, $A_1(-2,2)$ is the initial point and $A_2(-2,2)$ is the final point. The considered curve is closed.

Definition 1.25 The parameter t is called the parameter of the curve.

Definition 1.26 Equations (1.24) are called the parametric equations of the curve Γ.

Definition 1.27 We say that a curve Γ is an oriented curve in the sense that a point

$$(x_1', x_2', \ldots, x_m') = (\phi_1(t'), \phi_2(t'), \ldots \phi_m(t'))$$

is regarded as distinct from a point

$$(x_1'', x_2'', \ldots, x_m'') = (\phi_1(t''), \phi_2(t''), \ldots, \phi_m(t''))$$

if $t' \neq t''$ and as preceding $(x_1'', x_2'', \ldots, x_m'')$ if $t' < t''$. The oriented curve Γ is then said to be "traversed in the direction of increasing t."

Definition 1.28 Two curves Γ_1 and Γ_2 with equations

$$x_i = \phi_i(t), \quad i \in \{1, \ldots, m\}, \quad t \in I_1,$$

and

$$x_i = \psi_i(t), \quad i \in \{1, \ldots, m\}, \quad t \in I_2,$$

respectively, are regarded as identical if the equations of one curve can be transformed into the equations of the other curve by means of a continuous strictly increasing change of the parameter, i.e., if there is a continuous increasing function $\tau = \lambda(t), t \in [a,b]$, with the range $[\alpha, \gamma]$, such that

$$\psi_i(\lambda(t)) = \phi_i(t), \quad t \in [a,b].$$

We then say that the two curves have the same direction. We say that the two curves have opposite direction if the function λ is decreasing. In this case, the initial point of Γ_1 is the final point of Γ_2, and vice versa. The curve differing by Γ only by the direction in which it is traversed is denoted by $-\Gamma$.

Example 1.12 The curves

$$x_1 = t, \quad x_2 = t - 1, \quad x_3 = t^2 - 1, \quad t \in [0,1],$$

and

$$x_1 = t^2, \quad x_2 = t^2 - 1, \quad x_3 = t^4 - 1, \quad t \in [0,1],$$

are identical.

Definition 1.29 If the same point $(x_1, \ldots, x_m)$ corresponds to more than one parameter value in the half-open interval $[a,b)$, then we say that $(x_1, \ldots, x_m)$ is a multiple point of the curve (1.24).

Definition 1.30 A curve with no multiple points is called a simple curve or Jordan curve.

Example 1.13 The curve

$$x_1 = t, \quad x_2 = 4 - t, \quad x_3 = 5 + t, \quad t \in [1,4],$$

is a Jordan curve.

Let Γ be a continuous curve with equations (1.24). Consider a partition of the interval $[a,b]$,

$$P = \{t_0, t_1, \ldots, t_n\} \subset [a,b], \quad \text{where} \quad a = t_0 < t_1 < \ldots < t_n = b. \tag{1.25}$$

For a given partition P, we denote by $A_0, A_1, \ldots, A_n$ the corresponding points of the curve Γ, i.e., $A_j(\phi_1(t_j), \phi_2(t_j), \ldots, \phi_m(t_j))$, $j \in \{0, 1, \ldots, n\}$. We set

$$l(\Gamma, P) = \sum_{j=1}^{n} d(A_{j-1}, A_j) = \sum_{j=1}^{n} \sqrt{\sum_{k=1}^{m} \left(\phi_k(t_j) - \phi_k(t_{j-1})\right)^2}, \tag{1.26}$$

where $d(A_{j-1}, A_j)$ denotes the distance from point A_{j-1} to A_j.

Definition 1.31 The curve Γ is said to be rectifiable if

$$l(\Gamma) = \sup\{l(\Gamma, P) : P \quad \text{is a partition of} \quad [a,b]\} < \infty.$$

In this case, the nonnegative number $l = l(\Gamma)$ is called the length of the curve Γ. If the supremum does not exist, then the curve is said to be nonrectifiable.

Proposition 1.6 *If P and Q are partitions of $[a,b]$ and Q is a refinement of P, then*

$$l(\Gamma, P) \leq l(\Gamma, Q).$$

Theorem 1.40 *If the curve Γ is rectifiable, then its length does not depend on the parameterization of this curve.*

Theorem 1.41 *If a rectifiable curve Γ is split by means of a finite number of points $A_0, A_1, \ldots, A_n$ into a finite number of curves Γ_i and if the points A_i correspond to the values t_i of the parameter t and*

$$a = t_0 < t_1 < \ldots < t_n = b,$$

then each curve Γ_i is rectifiable and

$$l(\Gamma) = \sum_{i=1}^{n} l(\Gamma_i).$$

Theorem 1.42 *Assume the functions ϕ_i, $i \in \{1, \ldots, m\}$, are continuous on $[a,b]$ and β-differentiable on $[a,b)$. If their β-derivatives ϕ_i^{Δ} are bounded and β-integrable on $[a,b)$, then the curve Γ, defined by the parametric equations* (1.24), *is rectifiable, and its length $l(\Gamma)$ can be evaluated by the formula*

$$l(\Gamma) = \int_a^b \sqrt{\sum_{k=1}^{m} \left(\phi_k^{D_\beta}(t)\right)^2} D_\beta t.$$

Let Γ be a curve defined by equations (1.24). Put

$$A = (\phi_1(a), \ldots, \phi_m(a)) \quad \text{and} \quad B = (\phi_1(b), \ldots, \phi_m(b)).$$

Suppose that the function $f(x_1, \ldots, x_m)$ is defined and continuous on the curve Γ. Let

$$P = \{t_0, t_1, \ldots, t_n\} \subset [a,b],$$

where

$$a = t_0 < t_1 < \ldots < t_n = b,$$

be a partition of the interval $[a,b]$, and put

$$A_i = (\phi_1(t_i), \ldots, \phi_m(t_i)), \quad i \in \{0, 1, \ldots, n\}.$$

Take any $\tau_k \in [t_{k-1}, t_k)$, $k \in \{1, \ldots, n\}$. Denote by l_k the length of the piece of the curve Γ between its points A_{k-1} and A_k. Then, using Theorem 1.42, we have

$$l_k = \int_{t_{k-1}}^{t_k} \sqrt{\sum_{l=1}^{m} \left(\phi_l^{D_\beta}(t)\right)^2} D_\beta t.$$

We introduce

$$S_1 = \sum_{k=1}^{n} f(\phi_1(\tau_k), \ldots, \phi_m(\tau_k)) l_k.$$

Definition 1.32 We say that a complex number I_1 is the line integral of the first kind of the function f along the curve Γ if for each $\varepsilon > 0$, there exists $\delta > 0$ such that

$$|S_1 - I_1| < \varepsilon$$

for every integral sum S_1 of f corresponding to a partition $P \in \mathscr{P}_\delta([a,b])$ independent of the way in which $\tau_k \in [t_{k-1}, t_k)$ for $k \in \{1, 2, \ldots, n\}$ is chosen. We denote the number I_1 symbolically by

$$\int_\Gamma f(x_1, \ldots, x_m) D_\beta t \quad \text{or} \quad \int_{AB} f(x_1, \ldots, x_m) D_{\beta_l}, \tag{1.27}$$

where A and B are the initial and final points of the curve Γ, respectively.

Theorem 1.43 *Suppose that the curve Γ is given by the parametric equations* (1.24)*, where ϕ_i, $i \in \{1, \ldots, m\}$, are continuous on $[a,b]$ and β-differentiable on $[a,b)$. If ϕ_i^Δ, $i \in \{1, \ldots, m\}$, are bounded and β-integrable over $[a,b)$ and if the function f is continuous on Γ, then the line integral* (1.27) *exists and can be computed by the formula*

$$\int_\Gamma f(x_1; \ldots, x_m) D_{\beta_l} = \int_a^b f(\phi_1(t), \ldots, \phi_m(t)) \sqrt{\sum_{k=1}^{m} \left(\phi_k^{D_\beta}(t)\right)^2} D_\beta t.$$

Theorem 1.44 (Linearity)

Let $\alpha, \beta \in \mathbb{R}$. If the functions f and g are β-integrable along the curve Γ, then $\alpha f + \gamma g$ is also β-integrable along the curve Γ and

$$\int_\Gamma (\alpha f + \gamma g)(x_1, \ldots, x_m) D_{\beta_l} = \alpha \int_\Gamma f(x_1, \ldots, x_m) D_{\beta_l} + \gamma \int_\Gamma g(x_1, \ldots, x_m) D_{\beta_l}.$$

Theorem 1.45 (Additivity)

If the curve AB consists of two parts AC and CB and if the function f is β-integrable along the curve AB, then it is β-integrable along each of the curves AC and CB and

$$\int_{AB} f(x_1,\ldots,x_m)D_{\beta_l} = \int_{AC} f(x_1,\ldots,x_m)D_{\beta_l} + \int_{CB} f(x_1,\ldots,x_m)D_{\beta_l}.$$

Theorem 1.46 (Existence of Modulus of Integral)

If f is β-integrable along the curve Γ, then so is $|f|$ and

$$\left|\int_{\Gamma} f(x_1,\ldots,x_m)D_{\beta_l}\right| \leq \int_{\Gamma} |f(x_1,\ldots,x_m)|D_{\beta_l}. \tag{1.28}$$

Assume that g_i, $i \in \{1,\ldots,m\}$, are defined and continuous on the curve Γ. We introduce the integral sums

$$S_{2i} = \sum_{k=1}^{m} g_i(\phi_1(\tau_k),\ldots,\phi_m(\tau_k))(\phi_i(t_k) - \phi_i(t_{k-1})), \quad i = 1,\ldots,m.$$

Definition 1.33 We say that a complex number I_i, $i \in \{1,\ldots,m\}$, is the line integral of the second kind of the function g_i along the curve Γ if for each $\varepsilon > 0$, there exists $\delta > 0$ such that

$$|S_{2i} - I_i| < \varepsilon$$

for every integral sum S_{2i} of g_i corresponding to a partition

$$P \in \mathscr{P}_\delta([a,b])$$

independent of the way in which $\tau_k \in [t_{k-1}, t_k)$ for $k \in \{1,\ldots,n\}$ is chosen. We denote the number I_i symbolically by

$$\int_{\Gamma} g_i(x_1,\ldots,x_m)D_{\beta_i}x_i \quad \text{or} \quad \int_{AB} g_i(x_1,\ldots,x_m)D_{\beta_i}x_i. \tag{1.29}$$

The sum

$$\int_{\Gamma} g_1(x_1,\ldots,x_m)D_{\beta_1}x_1 + \cdots + \int_{\Gamma} g_m(x_1,\ldots,x_m)D_{\beta_m}x_m$$

is called a general line integral of the second kind, and it is denoted by

$$\int_{\Gamma} g_1(x_1,\ldots,x_m)D_{\beta_1}x_1 + \cdots + g_m(x_1,\ldots,x_m)D_{\beta_m}x_m.$$

Theorem 1.47 *Suppose that the curve Γ is given by the parametric equations (1.24), where ϕ_i are continuous on $[a,b]$ and β-integrable on $[a,b)$. If ϕ_i^{Δ} are bounded and β-integrable on $[a,b)$ and if the functions g_i are continuous on Γ, then the line integrals (1.29) exist and can be computed by the formula*

$$\int_{\Gamma} g_i(x_1,\ldots,x_m)D_{\beta_i}x_i = \int_a^b g_i(\phi_1(t),\ldots,\phi_m(t))\phi_i^{D_\beta}(t)D_\beta t.$$

Theorem 1.48 (Linearity)

Let $\alpha,\beta \in \mathbb{R}$. If f_i and g_i, $i \in \{1,\ldots,m\}$, are line β-integrable of the second kind along the curve Γ, then $\alpha f_i + \beta g_i$ is line β-integrable along the curve Γ and

$$\int_{\Gamma} (\alpha f_i + \beta g_i)(x_1,\ldots,x_m)D_{\beta_i}x_i$$

$$= \alpha \int_{\Gamma} f_i(x_1,\ldots,x_m)D_{\beta_i}x_i + \beta \int_{\Gamma} g_i(x_1,\ldots,x_m)D_{\beta_i}x_i.$$

Theorem 1.49 (Additivity)

If the curve AB consists of two curves AC and CB and if the functions g_i, $i \in \{1,\ldots,m\}$, are line β-integrable of the second kind along the curve AB, then they are line β-integrable of the second kind along each of the curves AC and CB and

$$\int_{AB} g_i(x_1,\ldots,x_m)D_{\beta_i}x_i = \int_{AC} g_i(x_1,\ldots,x_m)D_{\beta_i}x_i + \int_{CB} g_i(x_1,\ldots,x_m)D_{\beta_i}x_i.$$

1.3 The Green Formula

Let $I_1 \subset \mathbb{R}$ and $I_2 \subset \mathbb{R}$ be closed subsets. Let also, β_1 and β_2 be general first kind or second kind quantum operators in I_1 and I_2, respectively. Since I_1 and I_2 are closed subsets of $\mathbb{R}$, the set $I_1 \times I_2$ is a complete metric space with metric D defined by

$$d((x_1,x_2),(x_1',x_2')) = \sqrt{(x_1-x_1')^2+(x_2-x_2')^2}.$$

Definition 1.34 If $\mathscr{M}$ is a metric space, then any continuous mapping $h:[a,b] \to \mathscr{M}$ is called a continuous curve in $\mathscr{M}$.

Definition 1.35 Let $[a,b] \subset I_1$ with $a,b \in I_1$ and $y_0 \in I_2$. The set

$$\{(x,y_0) : x \in [a,b]\}$$

is called a horizontal line segment in $I_1 \times I_2$ and denoted by AB, where $A = (a, y_0)$ and $B = (b, y_0)$. We take $x_0 \in I_1$ and $[c,d] \subset I_2$ and define a vertical line segment in $I_1 \times I_2$ by

$$\{(x_0, y) : y \in [c,d]\}$$

and denote it by CD, where $C = (x_0, c)$ and $D = (x_0, d)$.

Definition 1.36 A finite sequence $A_1B_1, A_2B_2, \ldots, A_nB_n$, each of whose term A_kB_k, $k \in \{1, \ldots, n\}$, is a horizontal or vertical line segment in $I_1 \times I_2$, is said to form a polygonal path (or broken line) in $I_1 \times I_2$ with terminal points A_1 and B_n if

$$B_1 = A_2, \quad B_2 = A_3, \quad \ldots, \quad B_{n-1} = A_n.$$

Definition 1.37 A set of points of $I_1 \times I_2$ is said to be connected if any two of its points are terminal points of a polygonal path of points contained in the set.

Definition 1.38 A component of a set $\Omega \subset I_1 \times I_2$ is a nonempty maximal connected subset of Ω.

Definition 1.39 A nonempty open connected set of points of $I_1 \times I_2$ is called a domain.

Definition 1.40 A closed domain is a subset of $I_1 \times I_2$ which is the closure of a domain in $I_1 \times I_2$.

For the rectangle R in $I_1 \times I_2$ defined by

$$R = [a_1, b_1) \times [a_2, b_2), \tag{1.30}$$

we set

$$L_1 = \{(x, a_2) : x \in [a_1, b_1]\}, \quad L_2 = \{(b_1, y) : y \in [a_2, b_2]\},$$

$$L_3 = \{(x, b_2) : x \in [a_1, b_1]\}, \quad L_4 = \{(a_1, y) : y \in [a_2, b_2]\}.$$

Each of L_j, $j \in \{1,2,3,4\}$, is an oriented line segment.

Definition 1.41 The closed curve

$$\Gamma = L_1 \cup L_2 \cup (-L_3) \cup (-L_4) \tag{1.31}$$

is called the positively oriented fence of R, i.e., the rectangle Γ remains on the "left" side along the fence curve Γ.

Definition 1.42 We say that the set $E \subset I_1 \times I_2$ is a set of type ω if it is a connected set in $I_1 \times I_2$ which is the union of a finite number of rectangles of the form (1.30) that are pairwise disjoint and adjoining to each other.

Let $E \subset I_1 \times I_2$ be a set of type ω so that

$$E = \bigcup_{k=1}^{m} R_k,$$

where

$$R_k = [a_1^k, b_1^k) \times [a_2^k, b_2^k) \subset I_1 \times I_2, \quad k \in \{1, \ldots, m\},$$

and $R_1, R_2, \ldots, R_m$ are pairwise disjoint and adjoining to each other. Let Γ_k be the positively oriented fence of the rectangle R_k. We set

$$X = \bigcup_{k=1}^{m} \Gamma_k.$$

Let X_0 consist of a finite number of line segments each of which serves as a common part of fences of two adjoining rectangles belonging to $\{R_1, R_2, \ldots, R_m\}$.

Definition 1.43 The set

$$\Gamma = X \setminus X_0$$

forms a positively oriented closed "polygonal curve," which is called the positively oriented fence of the set E.

Theorem 1.50 (The Green Formula)

Let $E \subset I_1 \times I_2$ be a set of type ω and let Γ be its positively oriented fence. If the functions M and N are continuous and have continuous partial delta derivatives $M_{y_2} D_{\beta_2}$ and $N_{x_1} D_{\beta_1}$ on $E \cup \Gamma$, then

$$\int\int_E \left(N_{x_1} D_{\beta_1}(x_1, x_2) - M_{x_2} D_{\beta_2}(x_1, x_2) \right) D_{\beta_1} x_1 D_{\beta_2} x_2$$
$$= \int_\Gamma M(x_1, x_2) D_{\beta_1} x_1 + N(x_1, x_2) D_{\beta_2} x_2.$$

1.4 Advanced Practical Problems

Problem 1.1 Let $I_1 = \mathbb{N}_0 \cup \{-4\}$ and $I_2 = 4\mathbb{N}_0$. Let also,

$$\beta_1(t_1) = t_1 + 1, \quad t_1 \in I_1, \quad t_1 \neq 0,$$

$$\beta_2(t_2) = t_2 + 4, \quad t_2 \in I_2, \quad t_2 \neq 0,$$

$$\beta_1(0) = 0,$$

$$\beta_2(0) = 0.$$

Compute

$$\int_{-4}^{4}\int_{4}^{8} t_2 \log\frac{t_1+1}{t_1} D_{\beta_1}t_1 D_{\beta_2}t_2.$$

Problem 1.2 Let $\phi : [2,8] \to \mathbb{R}$, $\phi(x) = x^3$. Prove that ϕ satisfies the Lipschitz condition. Find a Lipschitz constant.

Problem 1.3 Let

$$\phi(x) = \begin{cases} \dfrac{1}{x^3-8} & \text{for} \quad x \in [0,2), \\ -1 & \text{for} \quad x = 2. \end{cases}$$

Check if ϕ satisfies the Lipschitz condition.

Problem 1.4 Let $I_1 = 2^{\mathbb{N}_0} \cup \{0\}$ and $I_2 = \mathbb{N}_0$. Let also,

$$\beta_1(t_1) = (\sqrt{t_1}+1)^2, \quad t_1 \in I_1, \quad t_1 \neq 0,$$

$$\beta_2(t_2) = t_2+1, \quad t_2 \in I_2, \quad t_2 \neq 0,$$

$$\beta_1(0) = 0,$$

$$\beta_2(0) = 0.$$

Prove that

$$\int_{-1}^{3}\int_{1}^{8} (t_1^3+3t_1^2t_2+t_1t_2^2+t_2^3) D_{\beta_1}t_1 D_{\beta_2}t_2 = \int_{-1}^{3} D_{\beta_2}t_2 \int_{1}^{8} (t_1^3+3t_1^2t_2+t_1t_2^2+t_2^3) D_{\beta_1}t_1$$

$$= \int_{1}^{8} D_{\beta_1}t_1 \int_{-1}^{3} (t_1^3+3t_1^2t_2+t_1t_2^2+t_2^3) D_{\beta_2}t_2.$$

Problem 1.5 Let

$$I_1 = I_2 = \mathbb{N} \cup \{0\}, \quad E = [1,2] \times [2,4].$$

Let also,

$$\beta_1(t_1) = t_1+1, \quad t_1 \in I_1, \quad t_1 \neq 0,$$

$$\beta_2(t_2) = t_2+1, \quad t_2 \in I_2, \quad t_2 \neq 0,$$

$$\beta_1(0) = 0,$$

$$\beta_2(0) = 0.$$

Find $J(E)$ and $\int\limits_E (2t_1 - t_2) D_{\beta_1} t_1 D_{\beta_2} t_2$.

Problem 1.6 Let $I_1 = I_2 = 2^{\mathbb{N}_0} \cup \{0\}$. Let also,

$$\beta_1(t_1) = (\sqrt{t_1} + 1)^2, \quad t_1 \in I_1, \quad t_1 \neq 0,$$

$$\beta_2(t_2) = (\sqrt{t_2} + 1)^2, \quad t_2 \in I_2, \quad t_2 \neq 0,$$

$$\beta_1(0) = 0,$$

$$\beta_2(0) = 0.$$

Compute the integral

$$\int\limits_1^3 \int\limits_{t_2}^{2t_2} (t_1^2 + t_1 t_2 + 3t_2^2) D_{\beta_1} t_1 D_{\beta_2} t_2.$$

Problem 1.7 Let $I_1 = I_2 = 4^{\mathbb{N}_0} \cup \{0\}$. Let also,

$$\beta_1(t_1) = (\sqrt[4]{t_1} + 1)^4, \quad t_1 \in I_1, \quad t_1 \neq 0,$$

$$\beta_2(t_2) = (\sqrt[4]{t_2} + 1)^4, \quad t_2 \in I_2, \quad t_2 \neq 0,$$

$$\beta_1(0) = 0,$$

$$\beta_2(0) = 0.$$

Compute the integral

$$\int\limits_0^4 \int\limits_0^{t_1} \left(\sinh_{f,\beta}(t_1) - 5t_2 \cosh_{f^2,\beta}(t_1) \right) D_{\beta_2} t_2 D_{\beta_1} t_1, \quad \text{where} \quad f(t_1) = t_1.$$

Problem 1.8 Let $I_1 = 2^{\mathbb{N}_0} \cup \{0\}$, $I_2 = 3^{\mathbb{N}_0} \cup \{0\}$. Let also,

$$\beta_1(t_1) = (\sqrt{t_1} + 1)^2, \quad t_1 \in I_1, \quad t_1 \neq 0,$$

$$\beta_2(t_2) = (\sqrt[3]{t_2} + 1)^3, \quad t_2 \in I_2, \quad t_2 \neq 0,$$

$$\beta_1(0) = 0,$$

$$\beta_2(0) = 0.$$

Compute the integral

$$\int_0^1 \int_0^{3t_1} (t_1 + 2t_2^2) D_{\beta_2} t_2 D_{\beta_1} t_1.$$

Problem 1.9 Let $I_1 = \mathbb{N}_0$ and $I_2 = 2^{\mathbb{N}_0} \cup \{0\}$. Let also,

$$\beta_1(t_1) = t_1 + 1, \quad t_1 \in I_1, \quad t_1 \neq 0,$$

$$\beta_2(t_2) = (\sqrt{t_2} + 1)^2, \quad t_2 \in I_2, \quad t_2 \neq 0,$$

$$\beta_1(0) = 0,$$

$$\beta_2(0) = 0.$$

Compute the integral

$$\int_0^2 \int_0^{t_1} \cosh_{f,\beta}(t_1)) D_{\beta_2} t_2 D_{\beta_1} t_1, \quad \text{where} \quad f(t_1) = t_1.$$

Problem 1.10 Let $I_1 = \mathbb{N}_0 \cup \{-1\}$ and $I_2 = 3^{\mathbb{N}_0} \cup \{0\}$. Let also,

$$\beta_1(t_1) = t_1 + 1, \quad t_1 \in I_1, \quad t_1 \neq -1,$$

$$\beta_2(t_2) = (\sqrt[3]{t_2} + 1)^3, \quad t_2 \in I_2, \quad t_2 \neq 0,$$

$$\beta_1(-1) = 0,$$

$$\beta_2(0) = 0.$$

Compute the integral

$$\int_{-1}^1 \int_0^{t_1} e_{f,\beta}(t_1) D_{\beta_2} t_2 D_{\beta_1} t_1, \quad \text{where} \quad f(t_1) = t_1.$$

Chapter 2
β-Differential Systems

In this chapter, we introduce the β-differential systems and we investigate their structures. We define β-exponential matrix-valued function and deduct some of its properties. The constant case is also considered. We prove the β-analogue of the classical Liouville theorem. The β-Putzer algorithm for finding the β-exponential matrix-valued function in the case when the matrix is constant is presented.

Suppose that $I \subseteq \mathbb{R}$ and $\beta : I \to \mathbb{R}$ is the first kind general quantum operator. Let s_0 be the unique fixed point of the operator β.

2.1 Structure of β-Differential Systems

Suppose that A is a $m \times n$-matrix on I, $A = (a_{ij})_{1 \le i \le m, 1 \le j \le n}$, shortly $A = (a_{ij})$, $a_{ij} : I \to \mathbb{R}$, $1 \le i \le m$, $1 \le j \le n$. A is also known as $m \times n$-matrix-valued function defined on I. The $m \times n$-identity matrix will be denoted by $\mathscr{I}$ and the $m \times n$-zero matrix by O.

Definition 2.1 We say that $m \times n$-matrix A is β-differentiable on I if each entry of A is β-differentiable on I and we write

$$D_\beta A = \left(D_\beta a_{ij}\right).$$

Example 2.1 Let $I = \mathbb{R}$ and define $\beta(t) = \frac{1}{2}t + 1$, $t \in I$, and

$$A(t) = \begin{pmatrix} t+1 & t^2+t \\ 2t-3 & 2t^2-3t+2 \end{pmatrix}, \quad t \in I.$$

We will find $D_\beta A(t)$, $t \in I$. We have

$$a_{11}(t) = t+1,$$

$$a_{12}(t) = t^2+t,$$

$$a_{21}(t) = 2t-3,$$

$$a_{22}(t) = 2t^2-3t+2, \quad t \in I.$$

Then

$$D_\beta a_{11}(t) = 1,$$

$$\begin{aligned} D_\beta a_{12}(t) &= \beta(t)+t+1 \\ &= \frac{1}{2}t+1+t+1 \\ &= \frac{3}{2}t+2, \end{aligned}$$

$$D_\beta a_{21}(t) = 2,$$

$$\begin{aligned} D_\beta a_{22}(t) &= 2(\beta(t)+t)-3 \\ &= 2\left(\frac{1}{2}t+1+t\right)-3 \\ &= 2\left(\frac{3}{2}t+1\right)-3 \\ &= 3t+2-3 \\ &= 3t-1, \quad t \in I. \end{aligned}$$

Therefore,

$$D_\beta A(t) = \begin{pmatrix} 1 & \frac{3}{2}t+2 \\ 2 & 3t-1 \end{pmatrix}, \quad t \in I.$$

Example 2.2 Let $I = \mathbb{R}$ and define

$$\beta(t) = \frac{2t}{3} + 2, \quad t \in I, \quad \text{and}$$

$$A(t) = \begin{pmatrix} t^3 + t & \frac{t+1}{t+2} & t \\ 1 & -t^2 & t^2 + t \\ 1 & 2 & t^3 \end{pmatrix}, \quad t \in I.$$

We will find $D_\beta A(t), t \in I$. We have

$$\begin{aligned} a_{11}(t) &= t^3 + t, \\ a_{12}(t) &= \frac{t+1}{t+2}, \\ a_{13}(t) &= t, \\ a_{21}(t) &= 1, \\ a_{22}(t) &= -t^2, \\ a_{23}(t) &= t^2 + t, \\ a_{31}(t) &= 1, \\ a_{32}(t) &= 2, \\ a_{33}(t) &= t^3, \quad t \in I. \end{aligned}$$

Then we obtain the following:

$$\begin{aligned} D_\beta a_{11}(t) &= (\beta(t))^2 + t\beta(t) + t^2 + 1 \\ &= \left(\frac{2}{3}t + 2\right)^2 + t\left(\frac{2}{3}t + 2\right) + t^2 + 1 \\ &= \frac{4}{9}t^2 + \frac{8}{3}t + 4 + \frac{2}{3}t^2 + 2t + t^2 + 1 \\ &= \frac{19}{9}t^2 + \frac{14}{3}t + 5, \end{aligned}$$

$$
\begin{aligned}
D_\beta a_{12}(t) &= \frac{t+2-(t+1)}{(t+2)(\beta(t)+2)} \\
&= \frac{1}{(t+2)\left(\frac{2}{3}t+2+2\right)} \\
&= \frac{3}{2(t+2)(t+6)}, \\
D_\beta a_{13}(t) &= 1, \\
D_\beta a_{21}(t) &= 0, \\
D_\beta a_{22}(t) &= -(\beta(t)+t) \\
&= -\left(\frac{2}{3}t+2+t\right) \\
&= -\left(\frac{5}{3}t+2\right), \\
D_\beta a_{23}(t) &= \beta(t)+t+1 \\
&= \frac{2}{3}t+2+t+1 \\
&= \frac{5}{3}t+3, \\
D_\beta a_{31}(t) &= 0, \\
D_\beta a_{32}(t) &= 0, \\
D_\beta a_{33}(t) &= (\beta(t))^2+t\beta(t)+t^2 \\
&= \left(\frac{2}{3}t+2\right)^2+t\left(\frac{2}{3}t+2\right)+t^2 \\
&= \frac{4}{9}t^2+\frac{8}{3}t+4+\frac{2}{3}t^2+2t+t^2 \\
&= \frac{19}{9}t^2+\frac{14}{3}t+4, \quad t\in I.
\end{aligned}
$$

Therefore,

$$D_\beta A(t) = \begin{pmatrix} \frac{19}{9}t^2 + \frac{14}{3}t + 5 & \frac{3}{2(t+2)(t+6)} & 1 \\ 0 & -\left(\frac{5}{3}t+2\right) & \frac{5}{3}t+3 \\ 0 & 0 & \frac{19}{9}t^2 + \frac{14}{3}t + 4 \end{pmatrix}, \quad t \in I.$$

Example 2.3 Let $I = \mathbb{R}$ and define $\beta(t) = \frac{3}{4}t + 1$, $t \in I$, and

$$A(t) = \begin{pmatrix} t^2+1 & \frac{1}{t+1} \\ 2 & 3t \end{pmatrix}, \quad t \in I.$$

We will find $D_\beta A(t)$, $t \in I$. We have

$$a_{11}(t) = t^2 + 1,$$

$$a_{12}(t) = \frac{1}{t+1},$$

$$a_{21}(t) = 2,$$

$$a_{22}(t) = 3t, \quad t \in I.$$

Then we obtain the following:

$$\begin{aligned} D_\beta a_{11}(t) &= \beta(t) + t \\ &= \frac{3}{4}t + 1 + t \\ &= \frac{7}{4}t + 1, \end{aligned}$$

$$\begin{aligned} D_\beta a_{12}(t) &= -\frac{1}{(t+1)(\beta(t)+1)} \\ &= -\frac{1}{(t+1)\left(\frac{3}{4}t+1+1\right)} \\ &= -\frac{4}{(t+1)(3t+8)}, \end{aligned}$$

$$D_\beta a_{21}(t) = 0,$$

$$D_\beta a_{22}(t) = 3, \quad t \in I.$$

Therefore,

$$D_\beta A(t) = \begin{pmatrix} \frac{7}{4}t+1 & -\frac{4}{(t+1)(3t+8)} \\ 0 & 3 \end{pmatrix}, \quad t \in I.$$

Exercise 2.1 Let $I = \mathbb{R}$ and define $\beta(t) = \frac{1}{2}t + 3$, $t \in I$, and

$$A(t) = \begin{pmatrix} t^3 & t^2 \\ 2t+4 & t-1 \end{pmatrix}, \quad t \in I.$$

Find $D_\beta A(t)$, $t \in I$.

Definition 2.2 If A is a β-differentiable $m \times n$-matrix, then we define

$$A^\beta = \left(a_{ij}^\beta\right).$$

Theorem 2.1 *If A is a β-differentiable $m \times n$-matrix-valued function on I, then*

$$A^\beta(t) = A(t) + (\beta(t) - t)D_\beta A(t), \quad t \in I.$$

Proof We have

$$\begin{aligned} A^\beta(t) &= \left(a_{ij}^\beta(t)\right) \\ &= \left(a_{ij}(t) + (\beta(t) - t)D_\beta a_{ij}(t)\right) \\ &= \left(a_{ij}(t)\right) + (\beta(t) - t)\left(D_\beta a_{ij}(t)\right) \\ &= A(t) + (\beta(t) - t)D_\beta A(t), \quad t \in I. \end{aligned}$$

This completes the proof. □

Below we suppose that $B = (b_{ij})_{1 \le i \le m, 1 \le j \le n}$, $b_{ij} : I \to \mathbb{R}$, $1 \le i \le m$, $1 \le j \le n$.

Theorem 2.2 *Let A and B be β-differentiable $m \times n$-matrix-valued functions on I. Then*

$$D_\beta(A + B) = D_\beta A + D_\beta B \quad \text{on} \quad I.$$

Proof We have

$$(A+B)(t) = (a_{ij}(t)+b_{ij}(t)),$$

$$D_\beta(A+B)(t) = \left(D_\beta a_{ij}(t)+D_\beta b_{ij}(t)\right)$$

$$= \left(D_\beta a_{ij}(t)\right)+\left(D_\beta b_{ij}(t)\right)$$

$$= D_\beta A(t)+D_\beta B(t), \quad t \in I.$$

This completes the proof. □

Theorem 2.3 *Let $\alpha \in \mathbb{R}$ and A be a β-differentiable $m \times n$-matrix-valued function on I. Then*

$$D_\beta(\alpha A) = \alpha D_\beta A \quad on \quad I.$$

Proof We have

$$D_\beta(\alpha A)(t) = \left(D_\beta(\alpha a_{ij})(t)\right)$$

$$= \left(\alpha D_\beta a_{ij}(t)\right)$$

$$= \alpha\left(D_\beta a_{ij}(t)\right)$$

$$= \alpha D_\beta A(t), \quad t \in I.$$

This completes the proof. □

Theorem 2.4 *Let A and B be two β-differentiable $n \times n$-matrix-valued functions on I. Then*

$$D_\beta(AB)(t) = D_\beta A(t)B(t)+A^\beta(t)D_\beta B(t)$$

$$= D_\beta A(t)B^\beta(t)+A(t)D_\beta B(t), \quad t \in I.$$

Proof We have

$$(AB)(t) = \left(\sum_{k=1}^{n} a_{ik}(t)b_{kj}(t)\right), \quad t \in I.$$

Then

$$\begin{aligned}
D_\beta(AB)(t) &= \left(D_\beta\left(\sum_{k=1}^{n} a_{ik}b_{kj}\right)(t)\right)\\
&= \left(\sum_{k=1}^{n} D_\beta(a_{ik}b_{kj})(t)\right)\\
&= \left(\sum_{k=1}^{n}\left(D_\beta a_{ik}(t)b_{kj}(t)+a_{ik}^{\beta}(t)D_\beta b_{kj}(t)\right)\right)\\
&= \left(\sum_{k=1}^{n} D_\beta a_{ik}(t)b_{kj}(t)\right)+\left(\sum_{k=1}^{n} a_{ik}^{\beta}(t)D_\beta b_{kj}(t)\right)\\
&= D_\beta A(t)B(t)+A^{\beta}(t)D_\beta B(t)\\
&= \left(\sum_{k=1}^{n}\left(D_\beta a_{ik}(t)b_{kj}^{\beta}(t)+a_{ik}(t)D_\beta b_{kj}(t)\right)\right)\\
&= \left(\sum_{k=1}^{n} D_\beta a_{ik}(t)b_{kj}^{\beta}(t)\right)+\left(\sum_{k=1}^{n} a_{ik}(t)D_\beta b_{kj}(t)\right)\\
&= D_\beta A(t)B^{\beta}(t)+A(t)D_\beta B(t).
\end{aligned}$$

Thus,

$$D_\beta(AB)(t) = D_\beta A(t)B^{\beta}(t)+A(t)D_\beta B(t), \quad t\in I.$$

This completes the proof. □

Example 2.4 Let $I=[0,\infty)$ and define $\beta(t)=\frac{1}{2}t+3$, $t\in I$,

$$A(t)=\begin{pmatrix} t & t-1\\ 2 & 3t+1\end{pmatrix}, \quad \text{and} \quad B(t)=\begin{pmatrix} 1 & t\\ t+1 & t-1\end{pmatrix}, \quad t\in I.$$

Then

$$\begin{aligned}
(AB)(t) &= \begin{pmatrix} t & t-1\\ 2 & 3t+1\end{pmatrix}\begin{pmatrix} 1 & t\\ t+1 & t-1\end{pmatrix}\\
&= \begin{pmatrix} t^2+t-1 & 2t^2-2t+1\\ 3t^2+4t+3 & 3t^2-1\end{pmatrix}
\end{aligned}$$

$$= C(t)$$

$$= (c_{ij}(t)), \quad t \in I.$$

We have

$$a_{11}(t) = t,$$

$$a_{12}(t) = t - 1,$$

$$a_{21}(t) = 2,$$

$$a_{22}(t) = 3t + 1,$$

$$b_{11}(t) = 1,$$

$$b_{12}(t) = t,$$

$$b_{21}(t) = t + 1,$$

$$b_{22}(t) = t - 1,$$

$$c_{11}(t) = t^2 + t - 1,$$

$$c_{12}(t) = 2t^2 - 2t + 1,$$

$$c_{21}(t) = 3t^2 + 4t + 3,$$

$$c_{22}(t) = 3t^2 - 1, \quad t \in I.$$

Then

$$D_\beta a_{11}(t) = 1,$$

$$D_\beta a_{12}(t) = 1,$$

$$D_\beta a_{21}(t) = 0,$$

$$
\begin{aligned}
D_\beta a_{22}(t) &= 3,\\
D_\beta b_{11}(t) &= 0,\\
D_\beta b_{12}(t) &= 1,\\
D_\beta b_{21}(t) &= 1,\\
D_\beta b_{22}(t) &= 1,\\
D_\beta c_{11}(t) &= \beta(t)+t+1\\
&= \frac{1}{2}t+3+t+1\\
&= \frac{3}{2}t+4,\\
D_\beta c_{12}(t) &= 2(\beta(t)+t)-2\\
&= 2\left(\frac{1}{2}t+3+t\right)-2\\
&= 3t+4,\\
D_\beta c_{21}(t) &= 3(\beta(t)+t)+4\\
&= 3\left(\frac{1}{2}t+3+t\right)+4\\
&= \frac{9}{2}t+13,\\
D_\beta c_{22}(t) &= 3(\beta(t)+t)\\
&= 3\left(\frac{1}{2}t+3+t\right)\\
&= \frac{9}{2}t+9, \quad t\in I.
\end{aligned}
$$

Therefore,

$$D_\beta(AB)(t) = \begin{pmatrix} \frac{3}{2}t+4 & 3t+4 \\ \frac{9}{2}t+13 & \frac{9}{2}t+9 \end{pmatrix}, \quad t \in I.$$

Also,

$$\begin{aligned} D_\beta A(t)B(t) &= \begin{pmatrix} 1 & 1 \\ 0 & 3 \end{pmatrix} \begin{pmatrix} 1 & t \\ t+1 & t-1 \end{pmatrix} \\ &= \begin{pmatrix} t+2 & 2t-1 \\ 3t+3 & 3t-3 \end{pmatrix}, \\ A^\beta(t) &= \begin{pmatrix} \beta(t) & \beta(t)-1 \\ 2 & 3\beta(t)+1 \end{pmatrix} \\ &= \begin{pmatrix} \frac{1}{2}t+3 & \frac{1}{2}t+2 \\ 2 & \frac{3}{2}t+10 \end{pmatrix}, \\ A^\beta(t)D_\beta B(t) &= \begin{pmatrix} \frac{1}{2}t+3 & \frac{1}{2}t+2 \\ 2 & \frac{3}{2}t+10 \end{pmatrix} \begin{pmatrix} 0 & 1 \\ 1 & 1 \end{pmatrix} \\ &= \begin{pmatrix} \frac{1}{2}t+2 & t+5 \\ \frac{3}{2}t+10 & \frac{3}{2}t+12 \end{pmatrix}, \\ D_\beta A(t)B(t) + A^\beta(t)D_\beta B(t) &= \begin{pmatrix} t+2 & 2t-1 \\ 3t+3 & 3t-3 \end{pmatrix} + \begin{pmatrix} \frac{1}{2}t+2 & t+5 \\ \frac{3}{2}t+10 & \frac{3}{2}t+12 \end{pmatrix} \\ &= \begin{pmatrix} \frac{3}{2}t+4 & 3t+4 \\ \frac{9}{2}t+13 & \frac{9}{2}t+9 \end{pmatrix}, \quad t \in I. \end{aligned}$$

Consequently,

$$D_\beta(AB)(t) = D_\beta A(t)B(t) + A^\beta(t)D_\beta B(t), \quad t \in I.$$

Next,

$$B^\beta(t) = \begin{pmatrix} 1 & \beta(t) \\ \beta(t)+1 & \beta(t)-1 \end{pmatrix} = \begin{pmatrix} 1 & \frac{1}{2}t+3 \\ \frac{1}{2}t+4 & \frac{1}{2}t+2 \end{pmatrix},$$

$$D_\beta A(t)B^\beta(t) = \begin{pmatrix} 1 & 1 \\ 0 & 3 \end{pmatrix} \begin{pmatrix} 1 & \frac{1}{2}t+3 \\ \frac{1}{2}t+4 & \frac{1}{2}t+2 \end{pmatrix} = \begin{pmatrix} \frac{1}{2}t+5 & t+5 \\ \frac{3}{2}t+12 & \frac{3}{2}t+6 \end{pmatrix},$$

$$A(t)D_\beta B(t) = \begin{pmatrix} t & t-1 \\ 2 & 3t+1 \end{pmatrix} \begin{pmatrix} 0 & 1 \\ 1 & 1 \end{pmatrix} = \begin{pmatrix} t-1 & 2t-1 \\ 3t+1 & 3t+3 \end{pmatrix},$$

$$D_\beta A(t)B^\beta(t) + A(t)D_\beta B(t) = \begin{pmatrix} \frac{1}{2}t+5 & t+5 \\ \frac{3}{2}t+12 & \frac{3}{2}t+6 \end{pmatrix} + \begin{pmatrix} t-1 & 2t-1 \\ 3t+1 & 3t+3 \end{pmatrix} = \begin{pmatrix} \frac{3}{2}t+4 & 3t+4 \\ \frac{9}{2}t+13 & \frac{9}{2}t+9 \end{pmatrix}, \quad t \in I.$$

Thus,

$$D_\beta(AB)(t) = D_\beta A(t)B^\beta(t) + A(t)D_\beta B(t), \quad t \in I.$$

Exercise 2.2 Let $I = \mathbb{R}$ and define $\beta(t) = \frac{1}{5}t$,

$$A(t) = \begin{pmatrix} t^2+1 & t-2 \\ 2t-1 & t+1 \end{pmatrix}, \quad \text{and}$$

$$B(t) = \begin{pmatrix} t & 2t+1 \\ t & t-1 \end{pmatrix}, \quad t \in I.$$

Then prove that

$$D_\beta(AB)(t) = D_\beta A(t) B^\beta(t) + A(t) D_\beta B(t), \quad t \in I.$$

Theorem 2.5 *Let A be an $n \times n$-matrix-valued function such that A^{-1} exists on I. Then*

$$\left(A^\beta\right)^{-1} = \left(A^{-1}\right)^\beta \quad \text{on} \quad I.$$

Proof For any $t \in I$, we have

$$A(t)A^{-1}(t) = \mathscr{I}, \quad t \in I.$$

Then

$$A^\beta(t)\left(A^{-1}\right)^\beta(t) = \mathscr{I},$$

whereupon

$$\left(A^\beta\right)^{-1}(t) = \left(A^{-1}\right)^\beta(t), \quad t \in I.$$

This completes the proof. □

Example 2.5 Let $I = \mathbb{R}$ and define $\beta(t) = lt + l$, $l \in (0, 1)$, and

$$A(t) = \begin{pmatrix} t+1 & t+2 \\ 1 & t+3 \end{pmatrix}, \quad t \in I.$$

Then

$$\begin{aligned} A^\beta(t) &= \begin{pmatrix} \beta(t)+1 & \beta(t)+2 \\ 1 & \beta(t)+3 \end{pmatrix} \\ &= \begin{pmatrix} lt+l+1 & lt+l+2 \\ 1 & lt+l+3 \end{pmatrix}, \\ (A^\beta)^{-1}(t) &= \frac{1}{(lt+l)(lt+l+3)+1} \begin{pmatrix} lt+l+3 & -lt-l-2 \\ -1 & lt+l+1 \end{pmatrix}, \quad t \in I. \end{aligned}$$

Next,

$$A^{-1}(t) = \frac{1}{t(t+3)+1}\begin{pmatrix} t+3 & -t-2 \\ -1 & t+1 \end{pmatrix}, \quad t \in I,$$

whereupon

$$\begin{aligned}\left(A^{-1}\right)^{\beta}(t) &= \frac{1}{\beta(t)(\beta(t)+3)+1}\begin{pmatrix} \beta(t)+3 & -\beta(t)-2 \\ -1 & \beta(t)+1 \end{pmatrix} \\ &= \frac{1}{(lt+l)(lt+l+3)+1}\begin{pmatrix} lt+l+3 & -lt-l-2 \\ -1 & lt+l+1 \end{pmatrix}, \quad t \in I.\end{aligned}$$

Consequently,

$$\left(A^{\beta}\right)^{-1}(t) = (A^{-1})^{\beta}(t), \quad t \in I.$$

Exercise 2.3 Let $I = \mathbb{R}$ and define $\beta(t) = \frac{2}{7}t + 4$, and

$$A(t) = \begin{pmatrix} t+2 & \frac{1}{t+1} \\ t^2+1 & \frac{1}{t+2} \end{pmatrix}, \quad t \in I.$$

Then prove that

$$(A^{\beta})^{-1}(t) = (A^{-1})^{\beta}(t), \quad t \in I.$$

Theorem 2.6 *Let A be a β-differentiable $n \times n$-matrix-valued function on I and A^{-1} such that $(A^{\beta})^{-1}$ exist on I. Then*

$$\begin{aligned} D_{\beta}\left(A^{-1}\right) &= -A^{-1}D_{\beta}A\left(A^{\beta}\right)^{-1} \\ &= -\left(A^{\beta}\right)^{-1}D_{\beta}AA^{-1} \quad \textit{on} \quad I. \end{aligned}$$

Proof We have

$$\mathscr{I} = AA^{-1} \quad on \quad I,$$

and

$$D_{\beta}(\mathscr{I}) = 0,$$

whereupon, using Theorem 2.4, we get

$$\begin{aligned} D_\beta(\mathscr{I}) &= D_\beta(AA^{-1}) \\ &= D_\beta A(A^{-1})^\beta + AD_\beta(A^{-1}) \\ &= D_\beta A(A^\beta)^{-1} + AD_\beta(A^{-1}) \\ &= D_\beta AA^{-1} + A^\beta D_\beta(A^{-1}) \quad on \quad I. \end{aligned}$$

Hence,

$$AD_\beta(A^{-1}) = -D_\beta A(A^\beta)^{-1},$$

$$A^\beta D_\beta(A^{-1}) = -D_\beta AA^{-1} \quad on \quad I,$$

and

$$D_\beta(A^{-1}) = -A^{-1}D_\beta A(A^\beta)^{-1},$$

$$D_\beta(A^{-1}) = -(A^\beta)^{-1}D_\beta AA^{-1} \quad on \quad I.$$

This completes the proof. □

Exercise 2.4 Let A and B be β-differentiable $n \times n$-matrix-valued functions on I such that B^{-1} and $(B^\beta)^{-1}$ exist on I. Then prove

$$\begin{aligned} D_\beta(AB^{-1}) &= \left(D_\beta A - AB^{-1}D_\beta B\right)(B^\beta)^{-1} \\ &= \left(D_\beta A - \left(AB^{-1}\right)^\beta D_\beta B\right)B^{-1} \quad on \quad I. \end{aligned}$$

Definition 2.3 We say that a matrix-valued function A is continuous on I if each entry of A is continuous. The class of such continuous $m \times n$-matrix-valued functions on I is denoted by

$$\mathscr{C} = \mathscr{C}(I) = \mathscr{C}(I, \mathscr{R}^{m\times n}).$$

Below, we suppose that A and B are $n \times n$-matrix-valued functions.

Definition 2.4 We say that an $n \times n$-matrix-valued function A on I is β-regressive with respect to I provided

$$\mathscr{I} + (\beta(t) - t)A(t) \quad \text{is invertible for all} \quad t \in I.$$

The class of such β-regressive and continuous functions is denoted, similar to the scalar case, by

$$\mathscr{R}_\beta = \mathscr{R}_\beta(I) = \mathscr{R}_\beta(I, \mathbb{R}^{n\times n}).$$

Theorem 2.7 *The $n \times n$-matrix-valued function A is β-regressive if and only if the eigenvalues $\lambda_i(t)$ of $A(t)$ are β-regressive for all $1 \le i \le n$.*

Proof Let $j \in \{1, \ldots, n\}$ be arbitrarily chosen and $\lambda_j(t)$ be an eigenvalue corresponding to the eigenvector $y(t)$. Then

$$\begin{aligned}(1+(\beta(t)-t)\lambda_j(t))y(t) &= \mathscr{I}y(t)+(\beta(t)-t)\lambda_j(t)y(t)\\ &= \mathscr{I}y(t)+(\beta(t)-t)A(t)y(t)\\ &= (\mathscr{I}+(\beta(t)-t)A(t))y(t),\end{aligned}$$

whereupon, it follows the assertion. This completes the proof. □

Example 2.6 Let $I = 3\mathbb{N}$ and define $\beta(t) = \frac{t}{2} + 3$ and

$$A(t) = \begin{pmatrix} 5t & t \\ 4t & 2t \end{pmatrix}, \quad t \in I.$$

Consider the equation

$$det \begin{pmatrix} 5t - \lambda(t) & t \\ 4t & 2t - \lambda(t) \end{pmatrix} = 0, \quad t \in I.$$

We have

$$(\lambda(t) - 5t)(\lambda(t) - 2t) - 4t^2 = 0, \quad t \in I,$$

or

$$(\lambda(t))^2 - 7t\lambda(t) + 10t^2 - 4t^2 = 0, \quad t \in I,$$

or

$$(\lambda(t))^2 - 7t\lambda(t) + 6t^2 = 0, \quad t \in I.$$

The roots of this equation are given by

$$\begin{aligned}\lambda_{1,2}(t) &= \frac{7t \pm \sqrt{49t^2 - 24t^2}}{2}\\ &= \frac{7t \pm 5t}{2}, \quad t \in I,\end{aligned}$$

or

$$\lambda_1(t) = 6t,$$

$$\lambda_2(t) = t, \quad t \in I.$$

Now,

$$1 + (\beta(t) - t)\lambda_{1,2}(t) = 0$$

implies

$$1 + \left(3 - \frac{t}{2}\right) 6t = 0$$

$$1 + \left(3 - \frac{t}{2}\right) t = 0, \quad t \in I,$$

or

$$3t^2 - 18t - 1 = 0$$

$$t^2 - 6t - 2 = 0, \quad t \in I.$$

The last two equations have no roots in I. Consequently,

$$1 + (\beta(t) - t)\lambda_{1,2}(t) \neq 0, \quad t \in I,$$

i.e., the matrix A is β-regressive.

Theorem 2.8 *Let A be a 2×2-matrix-valued function. Then A is β-regressive if and only if*

$$\operatorname{tr} A + (\beta(t) - t) \det A$$

is β-regressive. Here tr *A denotes the trace of the matrix A.*

Proof Let

$$A(t) = \begin{pmatrix} a_{11}(t) & a_{12}(t) \\ a_{21}(t) & a_{22}(t) \end{pmatrix}, \quad t \in I.$$

Then

$$\begin{aligned} \mathscr{I} + (\beta(t) - t)A(t) &= \begin{pmatrix} 1 & 0 \\ 0 & 1 \end{pmatrix} + \begin{pmatrix} (\beta(t) - t)a_{11}(t) & (\beta(t) - t)a_{12}(t) \\ (\beta(t) - t)a_{21}(t) & (\beta(t) - t)a_{22}(t) \end{pmatrix} \\ &= \begin{pmatrix} 1 + (\beta(t) - t)a_{11}(t) & (\beta(t) - t)a_{12}(t) \\ (\beta(t) - t)a_{21}(t) & 1 + (\beta(t) - t)a_{22}(t) \end{pmatrix}, \quad t \in I. \end{aligned}$$

Therefore, we obtain

$$\begin{aligned}\det(\mathscr{I}+(\beta(t)-t)A(t)) &= (1+(\beta(t)-t)a_{11}(t))(1+(\beta(t)-t)a_{22}(t)) \\ &\quad -((\beta(t)-t))^2 a_{12}(t)a_{21}(t) \\ &= 1+(\beta(t)-t)a_{22}(t)+(\beta(t)-t)a_{11}(t) \\ &\quad +((\beta(t)-t))^2 a_{11}(t)a_{22}(t)-((\beta(t)-t))^2 a_{12}(t)a_{21}(t) \\ &= 1+(\beta(t)-t)(\operatorname{tr} A)(t)+((\beta(t)-t))^2 (\det A)(t) \\ &= 1+(\beta(t)-t)\left((\operatorname{tr} A)(t)+(\beta(t)-t)(\det A)(t)\right). \end{aligned} \tag{2.1}$$

Suppose A is β-regressive. Then

$$\det(\mathscr{I}+(\beta(t)-t)A(t)) \neq 0, \quad t \in I.$$

In view of (2.1), we obtain

$$1+(\beta(t)-t)\left((\operatorname{tr} A)(t)+(\beta(t)-t)(\det A)(t)\right) \neq 0, \quad t \in I, \tag{2.2}$$

i.e.,

$$\operatorname{tr} A+(\beta(t)-t)\det A$$

is β-regressive. Conversely, suppose

$$\operatorname{tr} A+(\beta(t)-t)\det A$$

is β-regressive. Then (2.2) holds. From (2.1), we conclude that A is β-regressive. This completes the proof. □

Definition 2.5 Assume that A and B are β-regressive matrix-valued functions on I. Then we define $A \oplus_\beta B$, $\ominus_\beta A$, and $A \ominus_\beta B$ by

$$(A \oplus_\beta B)(t) = A(t)+B(t)+(\beta(t)-t)A(t)B(t),$$

$$(\ominus_\beta A)(t) = -(I+(\beta(t)-t)A(t))^{-1}A(t), \quad \text{and}$$

$$(A \ominus_\beta B)(t) = (A \oplus_\beta (\ominus_\beta B))(t), \quad t \in I,$$

respectively.

Example 2.7 Let $I = \mathbb{R}$ and define $\beta(t) = \frac{2}{3}t + 1$,

$$A(t) = \begin{pmatrix} 1 & t \\ 2 & 3t \end{pmatrix}, \quad \text{and}$$

$$B(t) = \begin{pmatrix} t & 1 \\ 2t & 3 \end{pmatrix}, \quad t \in I.$$

Here

$$\begin{aligned} \beta(t) - t &= \frac{2}{3}t + 1 - t \\ &= 1 - \frac{t}{3}, \quad t \in I. \end{aligned}$$

Then

$$\begin{aligned} (A \oplus_\beta B)(t) &= A(t) + B(t) + (\beta(t) - t)A(t)B(t) \\ &= \begin{pmatrix} 1 & t \\ 2 & 3t \end{pmatrix} + \begin{pmatrix} t & 1 \\ 2t & 3 \end{pmatrix} \\ &\quad + \left(1 - \frac{1}{3}t\right) \begin{pmatrix} 1 & t \\ 2 & 3t \end{pmatrix} \begin{pmatrix} t & 1 \\ 2t & 3 \end{pmatrix} \\ &= \begin{pmatrix} 1+t & 1+t \\ 2(1+t) & 3(1+t) \end{pmatrix} + \left(1 - \frac{1}{3}t\right) \begin{pmatrix} t + 2t^2 & 1 + 3t \\ 2t + 6t^2 & 2 + 9t \end{pmatrix} \\ &= \begin{pmatrix} -\frac{2}{3}t^3 + \frac{5}{3}t^2 + 2t + 1 & -t^2 + \frac{11}{3}t + 2 \\ -2t^3 + \frac{16}{3}t^2 + 4t + 2 & -3t^2 + \frac{34}{3}t + 5 \end{pmatrix}, \end{aligned}$$

and

$$
\begin{aligned}
(B \oplus_\beta A)(t) &= A(t)+B(t)+(\beta(t)-t)B(t)A(t)\\
&= \begin{pmatrix} 1 & t \\ 2 & 3t \end{pmatrix} + \begin{pmatrix} t & 1 \\ 2t & 3 \end{pmatrix}\\
&\quad + \left(1-\frac{1}{3}t\right)\begin{pmatrix} t & 1 \\ 2t & 3 \end{pmatrix}\begin{pmatrix} 1 & t \\ 2 & 3t \end{pmatrix}\\
&= \begin{pmatrix} 1+t & 1+t \\ 2(1+t) & 3(1+t) \end{pmatrix} + \left(1-\frac{1}{3}t\right)\begin{pmatrix} t+2 & t^2+3t \\ 2t+6 & 2t^2+9t \end{pmatrix}\\
&= \begin{pmatrix} -\frac{1}{3}t^2+\frac{4}{3}t+3 & -\frac{1}{3}t^3+4t+1 \\ -\frac{2}{3}t^2+2t+8 & -\frac{2}{3}t^3-t^2+12t+3 \end{pmatrix}.
\end{aligned}
$$

Also,

$$
\begin{aligned}
\mathscr{I}+(\beta(t)-t)B(t) &= \begin{pmatrix} 1 & 0 \\ 0 & 1 \end{pmatrix} + \left(1-\frac{1}{3}t\right)\begin{pmatrix} t & 1 \\ 2t & 3 \end{pmatrix}\\
&= \begin{pmatrix} -\frac{1}{3}t^2+t+1 & -\frac{1}{3}t+1 \\ -\frac{2}{3}t^2+2t & -t+4 \end{pmatrix},
\end{aligned}
$$

$$
\begin{aligned}
\det(\mathscr{I}+(\beta(t)-t)B(t)) &= \left(\frac{1}{3}t^2-t-1\right)(t-4)-\left(\frac{1}{3}t-1\right)\left(\frac{2}{3}t^2-2t\right)\\
&= \frac{1}{3}t^3-\frac{4}{3}t^2-t^2+4t-t+4-\left(\frac{2}{9}t^3-\frac{2}{3}t^2-\frac{2}{3}t^2+2t\right)\\
&= \frac{1}{3}t^3-\frac{7}{3}t^2+3t+4-\frac{2}{9}t^3+\frac{4}{3}t^2-2t\\
&= \frac{1}{9}t^3-t^2+t+4,
\end{aligned}
$$

and

$$(\mathscr{I}+(\beta(t)-t)B(t))^{-1} = \frac{1}{\frac{1}{9}t^3-t^2+t+4}\begin{pmatrix} -t+4 & -1+\frac{1}{3}t \\ -2t+\frac{2}{3}t^2 & -\frac{1}{3}t^2+t+1 \end{pmatrix},$$

$$\begin{aligned}(\ominus_\beta B)(t) &= -(\mathscr{I}+(\beta(t)-t)B(t))^{-1}B(t) \\ &= -\frac{1}{\frac{1}{9}t^3-t^2+t+4}\begin{pmatrix} -t+4 & -1+\frac{1}{3}t \\ -2t+\frac{2}{3}t^2 & -\frac{1}{3}t^2+t+1 \end{pmatrix}\begin{pmatrix} t & 1 \\ 2t & 3 \end{pmatrix} \\ &= -\frac{1}{\frac{1}{9}t^3-t^2+t+4}\begin{pmatrix} -\frac{1}{3}t^2+2t & 1 \\ 2t & -\frac{1}{3}t^2+t+3 \end{pmatrix}, \quad t \in I.\end{aligned}$$

Exercise 2.5 Let $I=\mathbb{R}$ and define $\beta(t)=2t$,

$$A(t) = \begin{pmatrix} 1 & 1 \\ 2 & -1 \end{pmatrix}, \quad \text{and}$$

$$B(t) = \begin{pmatrix} 3 & 4 \\ 1 & 0 \end{pmatrix}, \quad t \in I.$$

Then find

1. $(\ominus_\beta A)(t), t \in I$.
2. $(A \oplus_\beta B)(t), t \in I$.

Theorem 2.9 *The structure* $(\mathscr{R}_\beta, \oplus_\beta)$ *is a group.*

Proof Let $A, B, C \in (\mathscr{R}_\beta, \oplus_\beta)$. Then

$$(\mathscr{I}+(\beta(t)-t)A)^{-1}, \quad (\mathscr{I}+(\beta(t)-t)B)^{-1}, \quad (\mathscr{I}+(\beta(t)-t)C)^{-1}$$

exist and

$$\begin{aligned}\mathscr{I}+(\beta(t)-t)(A\oplus_\beta B) &= \mathscr{I}+(\beta(t)-t)(A+B+(\beta(t)-t)AB) \\ &= \mathscr{I}+(\beta(t)-t)A+(\beta(t)-t)B+(\beta(t)-t)^2AB \\ &= \mathscr{I}+(\beta(t)-t)A+(\mathscr{I}+(\beta(t)-t)A)(\beta(t)-t)B \\ &= (\mathscr{I}+(\beta(t)-t)A)(\mathscr{I}+(\beta(t)-t)B).\end{aligned}$$

Therefore,

$$(\mathscr{I}+(\beta(t)-t)(A\oplus_\beta B))^{-1}$$

exists. Also,

$$\begin{aligned} O\oplus_\beta A &= A\oplus_\beta O \\ &= A. \end{aligned}$$

Next,

$$\begin{aligned} A\oplus_\beta(-(\mathscr{I}+(\beta(t)-t)A)^{-1}A) &= A-(\mathscr{I}+(\beta(t)-t)A)^{-1}A \\ &\quad -(\beta(t)-t)(\mathscr{I}+(\beta(t)-t)A)^{-1}A^2 \\ &= A-(\mathscr{I}+(\beta(t)-t)A)^{-1}(\mathscr{I}+(\beta(t)-t)A)A \\ &= A-A \\ &= O, \end{aligned}$$

i.e., the additive inverse of A under the addition $\oplus_\beta$ is $-(\mathscr{I}+(\beta(t)-t)A)^{-1}A$. Note that

$$\begin{aligned} \mathscr{I}+(\beta(t)-t)(-(\mathscr{I}+(\beta(t)-t)A)^{-1}A) &= (\mathscr{I}+(\beta(t)-t)A)^{-1}(\mathscr{I}+(\beta(t)-t)A) \\ &\quad -(\mathscr{I}+(\beta(t)-t)A)^{-1}(\beta(t)-t)A \\ &= (\mathscr{I}+(\beta(t)-t)A)^{-1} \end{aligned}$$

and then $-(\mathscr{I}+(\beta(t)-t)A)^{-1}A\in\mathscr{R}_\beta$. Also,

$$\begin{aligned} (A\oplus_\beta B)\oplus_\beta C &= (A\oplus_\beta B)+C+(\beta(t)-t)(A\oplus_\beta B)C \\ &= A+B+(\beta(t)-t)AB+C+(\beta(t)-t)(A+B+(\beta(t)-t)AB)C \\ &= A+B+(\beta(t)-t)AB+C+(\beta(t)-t)AC+(\beta(t)-t)BC \\ &\quad +(\beta(t)-t)^2ABC, \end{aligned}$$

$$
\begin{aligned}
A \oplus_\beta (B \oplus_\beta C) &= A + (B \oplus_\beta C) + (\beta(t) - t)A(B \oplus_\beta C) \\
&= A + B + C + (\beta(t) - t)BC \\
&\quad + (\beta(t) - t)A(B + C + (\beta(t) - t)BC) \\
&= A + B + C + (\beta(t) - t)BC + (\beta(t) - t)AB + (\beta(t) - t)AC \\
&\quad + (\beta(t) - t)^2 ABC.
\end{aligned}
$$

Consequently,

$$(A \oplus_\beta B) \oplus_\beta C = A \oplus_\beta (B \oplus_\beta C),$$

i.e., in $(\mathscr{R}_\beta, \oplus_\beta)$ the associative law holds. This completes the proof. □

Henceforth, the conjugate of matrix A will be denoted by $\overline{A}$ and the transpose of matrix A will be denoted by A^T. The conjugate transpose of matrix of A will be denoted by $A^* = \left(\overline{A}\right)^T$.

Theorem 2.10 *Let A and B be β-regressive matrix-valued functions. Then we have the following:*

1. *A^* is regressive,*
2. *$\ominus_\beta A^* = \left(\ominus_\beta A\right)^*$,*
3. *$(A \oplus_\beta B)^* = B^* \oplus_\beta A^*$.*

Proof Since A is β-regressive, $(\mathscr{I} + (\beta(t) - t)A)^{-1}$ will exist.

1. We have

$$
\begin{aligned}
\mathscr{I} &= (\mathscr{I} + (\beta(t) - t)A)(\mathscr{I} + (\beta(t) - t)A)^{-1} \\
&= (\mathscr{I} + (\beta(t) - t)\overline{A})\overline{(\mathscr{I} + (\beta(t) - t)A)^{-1}} \\
&= \left((\mathscr{I} + (\beta(t) - t)A)^{-1}\right)^* (\mathscr{I} + (\beta(t) - t)A^*).
\end{aligned}
$$

Therefore,

$$(\mathscr{I} + (\beta(t) - t)A^*)^{-1} \quad \textit{and} \quad (\mathscr{I} + (\beta(t) - t)A^T)^{-1}$$

exist and

$$
\begin{aligned}
(\mathscr{I} + (\beta(t) - t)A^*)^{-1} &= \left((\mathscr{I} + (\beta(t) - t)A)^{-1}\right)^*, \\
(\mathscr{I} + (\beta(t) - t)A^T)^{-1} &= \left((\mathscr{I} + (\beta(t) - t)A)^{-1}\right)^T.
\end{aligned}
$$

Consequently, A^* is β-regressive.

2. We have

$$
\begin{aligned}
(\ominus_\beta A)^* &= -\left((\mathscr{I}+(\beta(t)-t)A)^{-1}A\right)^* \\
&= -A^*\left((\mathscr{I}+(\beta(t)-t)A)^{-1}\right)^* \\
&= -A^*(\mathscr{I}+(\beta(t)-t)A^*)^{-1} \\
&= \ominus_\beta A^*.
\end{aligned}
$$

Thus,

$$(\ominus_\beta A)^* = \ominus_\beta A^*.$$

3. We have

$$
\begin{aligned}
(A\oplus B)^* &= (A+B+(\beta(t)-t)AB)^* \\
&= A^*+B^*+(\beta(t)-t)B^*A^* \\
&= B^*+A^*+(\beta(t)-t)B^*A^* \\
&= B^*\oplus_\beta A^*.
\end{aligned}
$$

Thus,

$$(A\oplus B)^* = B^*\oplus_\beta A^*.$$

This completes the proof. □

2.2 β-Matrix Exponential Function

Definition 2.6 (β-Matrix Exponential Function) Let $A\in\mathscr{R}_\beta$ be a matrix-valued function and $t_0\in I$. The unique solution of the initial value problem (IVP)

$$D_\beta Y = A(t)Y, \quad t\in I,$$

$$Y(s_0) = \mathscr{I},$$

is called the β-matrix exponential function. It is denoted by $e_{A,\beta}(\cdot)$.

Theorem 2.11 *Let $A,B\in\mathscr{R}_\beta$ be matrix-valued functions and $t,s,r\in I$. Then we have the following:*

1. $e_{O,\beta}(t) = \mathscr{I}$, $e_{A_\beta}(s_0) = \mathscr{I}$,
2. $e_{A,\beta}(\beta(t)) = (\mathscr{I} + (\beta(t) - t)A(t))e_{A,\beta}(t)$,
3. $e_{A,\beta}(t)e_{B,\beta}(t) = e_{A\oplus_\beta B,\beta}(t)$, *if* $e_{A,\beta}(t)$ *and B commute.*

Proof 1. Consider the IVP

$$D_\beta Y = O, \quad t \in I,$$

$$Y(s_0) = \mathscr{I}.$$

Then, its unique solution is

$$e_{O,\beta}(t) = \mathscr{I}, \quad t \in I.$$

Now, we consider the IVP

$$D_\beta Y = A(t)Y, \quad t \in I,$$

$$Y(s_0) = \mathscr{I}.$$

By the definition of $e_{A,\beta}(\cdot)$, we obtain

$$e_{A,\beta}(s_0) = \mathscr{I}.$$

2. By Theorem 2.1 and the definition of $e_{A,\beta}(\cdot)$, we have

$$\begin{aligned} e_{A,\beta}(\beta(t)) &= e_{A,\beta}(t) + (\beta(t) - t)D_\beta e_{A,\beta}(t,s) \\ &= e_{A,\beta}(t) + (\beta(t) - t)A(t)e_{A,\beta}(t) \\ &= (\mathscr{I} + (\beta(t) - t)A(t))e_{A,\beta}(t). \end{aligned}$$

3. Let

$$Z(t) = e_{A,\beta}(t)e_{B,\beta}(t).$$

Then

$$\begin{aligned} D_\beta Z(t) &= D_\beta e_{A,\beta}(t)e^{\beta}_{B,\beta}(t) + e_{A,\beta}(t)D_\beta e_{B,\beta}(t) \\ &= A(t)e_{A,\beta}(t)(\mathscr{I} + (\beta(t) - t)B(t))e_{B,\beta}(t) \\ &\quad + B(t)e_{A,\beta}(t)e_{B,\beta}(t) \\ &= A(t)(\mathscr{I} + (\beta(t) - t)B(t))e_{A,\beta}(t)e_{B,\beta}(t) + B(t)e_{A,\beta}(t)e_{B,\beta}(t) \\ &= (A(t) + B(t) + (\beta(t) - t)A(t)B(t))e_{A,\beta}(t)e_{B,\beta}(t) \\ &= (A \oplus_\beta B)(t)Z(t). \end{aligned}$$

Also,

$$Z(s_0) = e_{A,\beta}(s_0)e_{B,\beta}(s_0)$$

$$= \mathscr{I}.$$

Consequently,

$$e_{A\oplus_\beta B,\beta}(t) = e_{A,\beta}(t)e_{B,\beta}(t).$$

This completes the proof. □

2.3 The β-Liouville Theorem

Theorem 2.12 (The β-Liouville Formula)
Let $A \in \mathscr{R}_\beta$ be a 2×2-matrix-valued function and assume that X is a solution of

$$D_\beta X = A(t)X.$$

Then X satisfies the β-Liouville formula

$$\det X(t) = e_{\operatorname{tr} A+(\beta(t)-t)\det A,\beta}(t)\det X(s_0), \quad t \in I.$$

Proof In view of Theorem 2.8, it follows that

$$\operatorname{tr} A + (\beta(t)-t)\det A$$

is β-regressive. Let

$$A(t) = \begin{pmatrix} a_{11}(t) & a_{12}(t) \\ a_{21}(t) & a_{22}(t) \end{pmatrix} \quad \text{and}$$

$$X(t) = \begin{pmatrix} x_{11}(t) & x_{12}(t) \\ x_{21}(t) & x_{22}(t) \end{pmatrix}.$$

Then

$$\begin{pmatrix} D_\beta x_{11}(t) & D_\beta x_{12}(t) \\ D_\beta x_{21}(t) & D_\beta x_{22}(t) \end{pmatrix} = \begin{pmatrix} a_{11}(t) & a_{12}(t) \\ a_{21}(t) & a_{22}(t) \end{pmatrix} \begin{pmatrix} x_{11}(t) & x_{12}(t) \\ x_{21}(t) & x_{22}(t) \end{pmatrix}$$

$$= \begin{pmatrix} a_{11}(t)x_{11}(t)+a_{12}(t)x_{12}(t) & a_{11}(t)x_{12}(t)+a_{12}(t)x_{22}(t) \\ a_{21}(t)x_{11}(t)+a_{22}(t)x_{21}(t) & a_{21}(t)x_{12}(t)+a_{22}(t)x_{22}(t) \end{pmatrix},$$

whereupon

$$\begin{cases} D_\beta x_{11}(t) &= a_{11}(t)x_{11}(t) + a_{12}(t)x_{21}(t) \\ D_\beta x_{12}(t) &= a_{11}(t)x_{12}(t) + a_{12}(t)x_{22}(t) \\ D_\beta x_{21}(t) &= a_{21}(t)x_{11}(t) + a_{22}(t)x_{21}(t) \\ D_\beta x_{22}(t) &= a_{21}(t)x_{12}(t) + a_{22}(t)x_{22}(t). \end{cases}$$

Then

$$\det X(t) = x_{11}(t)x_{22}(t) - x_{12}(t)x_{21}(t) \quad \text{and}$$

$$\begin{aligned} D_\beta(\det X)(t) &= D_\beta x_{11}(t)x_{22}(t) + x_{11}^\beta(t)D_\beta x_{22}(t) \\ &\quad - D_\beta x_{12}(t)x_{21}(t) - x_{12}^\beta(t)D_\beta x_{21}(t) \\ &= \det\begin{pmatrix} D_\beta x_{11}(t) & D_\beta x_{12}(t) \\ x_{21}(t) & x_{22}(t) \end{pmatrix} + \det\begin{pmatrix} x_{11}^\beta(t) & x_{12}^\beta(t) \\ D_\beta x_{21}(t) & D_\beta x_{22}(t) \end{pmatrix} \\ &= \det\begin{pmatrix} a_{11}(t)x_{11}(t) + a_{12}(t)x_{21}(t) & a_{11}(t)x_{12}(t) + a_{12}(t)x_{22}(t) \\ x_{21}(t) & x_{22}(t) \end{pmatrix} \\ &\quad + \det\begin{pmatrix} x_{11}(t) + (\beta(t)-t)D_\beta x_{11}(t) & x_{12}(t) + (\beta(t)-t)D_\beta x_{12}(t) \\ D_\beta x_{21}(t) & D_\beta x_{22}(t) \end{pmatrix} \\ &= [a_{11}(t)x_{11}(t)x_{22}(t) + a_{12}(t)x_{21}(t)x_{22}(t) \\ &\quad - a_{11}(t)x_{12}(t)x_{21}(t) - a_{12}(t)x_{21}(t)x_{22}(t)] \\ &\quad + \det\begin{pmatrix} x_{11}(t) & x_{12}(t) \\ D_\beta x_{21}(t) & x_{22}(t) \end{pmatrix} \\ &\quad + (\beta(t)-t)(t)\det\begin{pmatrix} D_\beta x_{11}(t) & D_\beta x_{12}(t) \\ D_\beta x_{21}(t) & D_\beta x_{22}(t) \end{pmatrix} \\ &= a_{11}(t)\det X(t) + \det\begin{pmatrix} x_{11}(t) & x_{12}(t) \\ D_\beta x_{21}(t) & D_\beta x_{22}(t) \end{pmatrix} \\ &\quad + (\beta(t)-t)\det\begin{pmatrix} D_\beta x_{11}(t) & D_\beta x_{12}(t) \\ D_\beta x_{21}(t) & D_\beta x_{22}(t) \end{pmatrix} \end{aligned}$$

$$
\begin{aligned}
&= a_{11}(t)\det X(t) \\
&\quad + \det \begin{pmatrix} x_{11}(t) & x_{12}(t) \\ a_{21}(t)x_{11}(t) + a_{22}(t)x_{21}(t) & a_{21}(t)x_{12}(t) + a_{22}(t)x_{22}(t) \end{pmatrix} \\
&\quad + (\beta(t) - t) \det \begin{pmatrix} a_{11}(t)x_{11}(t) + a_{12}(t)x_{21}(t) & a_{11}(t)x_{12}(t) + a_{12}(t)x_{22}(t) \\ D_\beta x_{21}(t) & D_\beta x_{22}(t) \end{pmatrix} \\
&= a_{11}(t)\det X(t) + [a_{21}(t)x_{11}(t)x_{12}(t) + a_{22}(t)x_{11}(t)x_{22}(t) \\
&\quad - a_{21}(t)x_{11}(t)x_{12}(t) - a_{22}(t)x_{21}(t)x_{12}(t)] \\
&\quad + (\beta(t) - t) \left[a_{11}(t) \det \begin{pmatrix} x_{11}(t) & x_{12}(t) \\ D_\beta x_{21}(t) & D_\beta x_{22}(t) \end{pmatrix} \right. \\
&\quad \left. + a_{12}(t) \det \begin{pmatrix} x_{21}(t) & x_{22}(t) \\ D_\beta x_{21}(t) & D_\beta x_{22}(t) \end{pmatrix} \right] \\
&= a_{11}(t)\det X(t) + a_{22}(t)\det X(t) \\
&\quad + (\beta(t) - t) \left[a_{11}(t) \det \begin{pmatrix} x_{11}(t) & x_{12}(t) \\ a_{21}(t)x_{11}(t) + a_{22}(t)x_{21}(t) & a_{21}(t)x_{12}(t) + a_{22}(t)x_{22}(t) \end{pmatrix} \right. \\
&\quad \left. + a_{12}(t) \det \begin{pmatrix} x_{21}(t) & x_{22}(t) \\ a_{21}(t)x_{11}(t) + a_{22}(t)x_{21}(t) & a_{21}(t)x_{12}(t) + a_{22}(t)x_{22}(t) \end{pmatrix} \right] \\
&= \operatorname{tr} A(t)\det X(t) \\
&\quad + (\beta(t) - t) [a_{11}(t)(a_{21}(t)x_{12}(t)x_{11}(t) + a_{22}(t)x_{22}(t)x_{11}(t) \\
&\quad - a_{21}(t)x_{11}(t)x_{12}(t) - a_{22}(t)x_{12}(t)x_{21}(t)) \\
&\quad + a_{12}(t)(a_{21}(t)x_{21}(t)x_{12}(t) + a_{22}(t)x_{22}(t)x_{21}(t) \\
&\quad - a_{21}(t)x_{11}(t)x_{22}(t) - a_{22}(t)x_{21}(t)x_{22}(t))] \\
&= \operatorname{tr} A(t)\det X(t) + (\beta(t) - t)[a_{11}(t)a_{22}(t)\det X(t) - a_{12}(t)a_{21}(t)\det X(t)] \\
&= \operatorname{tr} A(t)\det X(t) + (\beta(t) - t)\det A(t)\det X(t) \\
&= [\operatorname{tr} A(t) + (\beta(t) - t)\det A(t)]\det X(t),
\end{aligned}
$$

i.e.,

$$D_\beta(\det X)(t) = [\operatorname{tr} A(t) + (\beta(t) - t)\det A(t)]\det X(t).$$

Thus,

$$\det X(t) = e_{\operatorname{tr} A + (\beta(t)-t)\det A}(t)\det X(s_0).$$

This completes the proof. $\square$

Example 2.8 Let $I = \mathbb{R}$ and define $\beta(t) = \frac{1}{2}t + 1$, $t \in I$. Here $s_0 = 2$ Consider the IVP

$$D_\beta X(t) = \begin{pmatrix} 3t & 4t+1 \\ 3 & 2+t \end{pmatrix} X(t), \quad t \in I,$$

$$X(2) = \begin{pmatrix} 1 & 0 \\ 1 & 1 \end{pmatrix}.$$

Here

$$A(t) = \begin{pmatrix} 3t & 4t+1 \\ 3 & 2+t \end{pmatrix}$$

$$\beta(t) - t = 1 - \frac{1}{2}t, \quad t \in I.$$

Then

$$\begin{aligned} \det A(t) &= 3t(2+t) - 3(4t+1) \\ &= 6t + 3t^2 - 12t - 3 \\ &= 3t^2 - 6t - 3, \end{aligned}$$

$$\begin{aligned} \operatorname{tr} A(t) &= 3t + 2 + t \\ &= 2 + 4t, \end{aligned}$$

$$\begin{aligned} \operatorname{tr} A(t) + (\beta(t) - t)\det A(t) &= 2 + 4t + \left(1 - \frac{1}{2}t\right)(3t^2 - 6t - 3) \\ &= 2 + 4t + 3t^2 - 6t - 3 - \frac{3}{2}t^3 + 3t^2 + \frac{3}{2}t \\ &= -\frac{3}{2}t^3 + 6t^2 - \frac{1}{2}t - 1, \quad t \in I. \end{aligned}$$

Let

$$f(t) = -\frac{3}{2}t^3 + 6t^2 - \frac{1}{2}t - 1, \quad t \in I.$$

Note that

$$\beta^k(t) = \left(\frac{1}{2}\right)^k t + [k]_{\frac{1}{2}}, \quad k \in \mathbb{N}.$$

Then, using the β-Liouville formula, we get

$$\begin{aligned}\det X(t) &= e_{f,\beta}(t) \det X(2) \\ &= e_{f,\beta}(t) \\ &= \frac{1}{\prod\limits_{k=0}^{\in fty} (1 - f(\beta^k(t))(\beta^k(t) - \beta^{k+1}(t)))}, \quad t \in I.\end{aligned}$$

Exercise 2.6 Let $A \in \mathscr{R}_\beta$ be a $n \times n$-matrix-valued function and assume that X is a solution of

$$D_\beta X = A(t)X.$$

Then prove that X satisfies the β-Liouville formula

$$\det X(t) = e_{\operatorname{tr} A + (\beta(t)-t)\det A, \beta}(t) \det X(s_0), \quad t \in I.$$

Hint Use Theorem 2.12 and the principle of the mathematical induction. □

2.4 Constant Coefficients

In this section, we suppose that A is an $n \times n$-matrix-valued constant function such that $A \in \mathscr{R}_\beta$. Consider the β-differential system

$$D_\beta x = Ax. \tag{2.3}$$

Theorem 2.14 *Let λ and ξ be an eigenpair of A. Then*

$$x(t) = e_{\lambda,\beta}(t)\xi, \quad t \in I,$$

is a solution of (2.3).

Proof Since

$$A\xi = \lambda\xi.$$

For $t \in I$, we have

$$\begin{aligned} D_\beta x(t) &= D_\beta e_{\lambda,\beta}(t)\xi \\ &= \lambda e_{\lambda,\beta}(t)\xi \\ &= e_{\lambda,\beta}(t)(\lambda\xi) \\ &= e_{\lambda,\beta}(t)A\xi \\ &= A\left(e_\lambda(t,t_0)\xi\right) \\ &= Ax(t). \end{aligned}$$

Thus,

$$D_\beta x(t) = Ax(t), \quad t \in I.$$

That is, x is a solution of (2.3). This completes the proof. □

Example 2.9 Let $I \subseteq \mathbb{R}$ and consider the β-differential system

$$\begin{aligned} D_\beta x_1(t) &= -3x_1 - 2x_2 \\ D_\beta x_2(t) &= 3x_1 + 4x_2, \quad t \in I. \end{aligned}$$

Here

$$A = \begin{pmatrix} -3 & -2 \\ 3 & 4 \end{pmatrix}.$$

Then

$$\det\begin{pmatrix} -3-\lambda & -2 \\ 3 & 4-\lambda \end{pmatrix} = 0$$

implies

$$\lambda^2 - \lambda - 6 = 0,$$

which yields

$$\begin{aligned} \lambda_1 &= 3, \\ \lambda_2 &= -2. \end{aligned}$$

The considered β-differential system is β-regressive for $I \subseteq \mathbb{R}$ for which $-2 \in \mathscr{R}_\beta$. Note that

$$\xi_1 = \begin{pmatrix} 1 \\ \\ -3 \end{pmatrix},$$

$$\xi_2 = \begin{pmatrix} -2 \\ \\ 1 \end{pmatrix}$$

are eigenvalues corresponding to λ_1 and λ_2, respectively. Therefore,

$$x(t) = c_1 e_{3,\beta}(t)\xi_1 + c_2 e_{-2,\beta}(t)\xi_2$$

$$= c_1 e_{3,\beta}(t) \begin{pmatrix} 1 \\ \\ -3 \end{pmatrix} + c_2 e_{-2,\beta}(t) \begin{pmatrix} -2 \\ \\ 1 \end{pmatrix}, \quad t \in I,$$

where c_1 and c_2 are real constants, is a general solution of the considered β-differential system, if $-2 \in \mathscr{R}_\beta$.

Example 2.10 Let $I \subseteq \mathbb{R}$ and consider the β-differential system

$$D_\beta x_1(t) = x_1(t) - x_2(t)$$

$$D_\beta x_2(t) = -x_1(t) + 2x_2(t) - x_3(t)$$

$$D_\beta x_3(t) = -x_2(t) + x_3(t), \quad t \in I.$$

Here

$$A = \begin{pmatrix} 1 & -1 & 0 \\ \\ -1 & 2 & -1 \\ \\ 0 & -1 & 1 \end{pmatrix}.$$

Then

$$\det(A - \lambda \mathscr{I}) = \det \begin{pmatrix} 1-\lambda & -1 & 0 \\ \\ -1 & 2-\lambda & -1 \\ \\ 0 & -1 & 1-\lambda \end{pmatrix}$$

$$= -(\lambda-1)^2(\lambda-2) + (\lambda-1) + (\lambda-1)$$

$$= (\lambda-1)(-(\lambda-1)(\lambda-2) + 2)$$

$$= (\lambda - 1)\left(-\lambda^2 + 3\lambda\right)$$

$$= -\lambda(\lambda - 1)(\lambda - 3),$$

whereupon

$$\det(A - \lambda \mathscr{I}) = 0$$

yields

$$\lambda_1 = 0,$$

$$\lambda_2 = 1,$$

$$\lambda_3 = 3.$$

Note that the matrix A is β-regressive and

$$\xi_1 = \begin{pmatrix} 1 \\ 1 \\ 1 \end{pmatrix},$$

$$\xi_2 = \begin{pmatrix} 1 \\ 0 \\ -1 \end{pmatrix}, \quad \text{and}$$

$$\xi_3 = \begin{pmatrix} 1 \\ -2 \\ 1 \end{pmatrix}$$

are eigenvalues corresponding to λ_1, λ_2, and λ_3, respectively. Consequently,

$$x(t) = c_1\xi_1 + c_2 e_{1,\beta}(t)\xi_2 + c_3 e_{3,\beta}(t)\xi_3$$

$$= c_1 \begin{pmatrix} 1 \\ 1 \\ 1 \end{pmatrix} + c_2 e_{1,\beta}(t) \begin{pmatrix} 1 \\ 0 \\ -1 \end{pmatrix} + c_3 e_{3,\beta}(t) \begin{pmatrix} 1 \\ -2 \\ 1 \end{pmatrix}, \quad t \in I,$$

where c_1, c_2, and c_3 are constants, is a general solution of the considered β-differential system.

Example 2.11 Let $I \subseteq \mathbb{R}$, and consider the β-differential system

$$D_\beta x_1(t) = -x_1(t) + x_2(t) + x_3(t)$$

$$D_\beta x_2(t) = x_2(t) - x_3(t) + x_4(t)$$

$$D_\beta x_3(t) = 2x_3(t) - 2x_4(t)$$

$$D_\beta x_4(t) = 3x_4(t), \quad t \in I.$$

Here

$$A = \begin{pmatrix} -1 & 1 & 1 & 0 \\ 0 & 1 & -1 & 1 \\ 0 & 0 & 2 & -2 \\ 0 & 0 & 0 & 3 \end{pmatrix}.$$

Then

$$\det(A - \lambda \mathscr{I}) = \det \begin{pmatrix} -1-\lambda & 1 & 1 & 0 \\ 0 & 1-\lambda & -1 & 1 \\ 0 & 0 & 2-\lambda & -2 \\ 0 & 0 & 0 & 3-\lambda \end{pmatrix}$$

$$= (\lambda+1)(\lambda-1)(\lambda-2)(\lambda-3),$$

whereupon

$$\det(A - \lambda \mathscr{I}) = 0$$

yields

$$\lambda_1 = -1,$$

$$\lambda_2 = 1,$$

$$\lambda_3 = 2,$$

$$\lambda_4 = 3.$$

The matrix A is β-regressive, if $-1 \in \mathscr{R}_\beta$. Note that

$$\xi_1 = \begin{pmatrix} 1 \\ 0 \\ 0 \\ 0 \end{pmatrix},$$

$$\xi_2 = \begin{pmatrix} 0 \\ 1 \\ 0 \\ 0 \end{pmatrix},$$

$$\xi_3 = \begin{pmatrix} 0 \\ 0 \\ 1 \\ 0 \end{pmatrix}, \quad \text{and}$$

$$\xi_4 = \begin{pmatrix} 0 \\ 0 \\ 0 \\ 1 \end{pmatrix}$$

are eigenvectors corresponding to λ_1, λ_2, λ_3, and λ_4, respectively. Consequently,

$$\begin{aligned} x(t) &= c_1 e_{-1,\beta}(t)\xi_1 + c_2 e_{1,\beta}(t)\xi_2 + c_3 e_{2,\beta}(t)\xi_3 + c_4 e_{5,\beta}(t)\xi_4 \\ &= c_1 e_{-1,\beta}(t) \begin{pmatrix} 1 \\ 0 \\ 0 \\ 0 \end{pmatrix} + c_2 e_{1,\beta}(t) \begin{pmatrix} 0 \\ 1 \\ 0 \\ 0 \end{pmatrix} \\ &\quad + c_3 e_{2,\beta}(t) \begin{pmatrix} 0 \\ 0 \\ 1 \\ 0 \end{pmatrix} + c_4 e_{3,\beta}(t) \begin{pmatrix} 0 \\ 0 \\ 0 \\ 1 \end{pmatrix}, \quad t \in I, \end{aligned}$$

where c_1, c_2, c_3, and c_4 are real constants, is a general solution of the considered β-differential system.

Exercise 2.7 Find a general solution of the β-differential system

$$D_\beta x_1(t) = x_2(t)$$

$$D_\beta x_2(t) = x_1(t), \quad t \in I.$$

Theorem 2.15 *Assume that* $A \in \mathscr{R}_\beta$. *If*

$$x(t) = u(t) + iv(t), \quad t \in I,$$

is a complex vector-valued solution of (2.3), *where u and v are real vector-valued functions on I, then u and v are real vector-valued solutions of* (2.3) *on I.*

Proof We have

$$\begin{aligned} D_\beta x(t) &= A(t)x(t) \\ &= A(t)\left(u(t) + iv(t)\right) \\ &= A(t)u(t) + iA(t)v(t) \\ &= D_\beta u(t) + iD_\beta v(t), \quad t \in I. \end{aligned}$$

Equating real and imaginary parts, we get

$$D_\beta u(t) = A(t)u(t),$$

$$D_\beta v(t) = A(t)v(t), \quad t \in I.$$

This means that u and v are solutions of (2.3) on I. This completes the proof. □

Example 2.12 Consider the β-differential system

$$D_\beta x_1(t) = x_1(t) + x_2(t)$$

$$D_\beta x_2(t) = -x_1(t) + x_2(t), \quad t \in I.$$

Here

$$A = \begin{pmatrix} 1 & 1 \\ -1 & 1 \end{pmatrix}.$$

Then

$$\begin{aligned}\det(A-\lambda\mathscr{I}) &= \det\begin{pmatrix}1-\lambda & 1\\ -1 & 1-\lambda\end{pmatrix}\\ &= (\lambda-1)^2+1\\ &= \lambda^2-2\lambda+1+1\\ &= \lambda^2-2\lambda+2,\end{aligned}$$

whereupon

$$\det(A-\lambda\mathscr{I})=0$$

yields

$$\lambda_{1,2}=1\pm i.$$

Note that

$$\xi=\begin{pmatrix}1\\ i\end{pmatrix}$$

is an eigenvector corresponding to the eigenvalue $\lambda=1+i$. We have

$$\begin{aligned}x(t) &= e_{1+i,\beta}(t)\begin{pmatrix}1\\ i\end{pmatrix}\\ &= e_{1,\beta}(t)\left(\cos_{\frac{1}{1+(\beta(t)-t)},\beta}(t)+i\sin_{\frac{1}{1+(\beta(t)-t)},\beta}(t)\right)\begin{pmatrix}1\\ i\end{pmatrix}\\ &= e_{1,\beta}(t)\left(\begin{pmatrix}\cos_{\frac{1}{1+(\beta(t)-t)},\beta}(t)\\ i\cos_{\frac{1}{1+(\beta(t)-t)},\beta}(t)\end{pmatrix}+\begin{pmatrix}i\sin_{\frac{1}{1+(\beta(t)-t)},\beta}(t)\\ -\sin_{\frac{1}{1+(\beta(t)-t)},\beta}(t)\end{pmatrix}\right)\\ &= e_{1,\beta}(t)\begin{pmatrix}\cos_{\frac{1}{1(\beta(t)-t)},\beta}(t)\\ -\sin_{\frac{1}{1(\beta(t)-t)},\beta}(t)\end{pmatrix}+ie_{1,\beta}(t)\begin{pmatrix}\sin_{\frac{1}{1+(\beta(t)-t)},\beta}(t)\\ \cos_{\frac{1}{1+(\beta(t)-t)},\beta}(t)\end{pmatrix}.\end{aligned}$$

Consequently,

$$e_{1,\beta}(t)\begin{pmatrix}\cos_{\frac{1}{1+(\beta(t)-t)},\beta}(t)\\ -\sin_{\frac{1}{1+(\beta(t)-t)},\beta}(t)\end{pmatrix}\quad \textit{and}\quad e_{1,\beta}(t)\begin{pmatrix}\sin_{\frac{1}{1+(\beta(t)-t)},\beta}(t)\\ \cos_{\frac{1}{1+(\beta(t)-t)},\beta}(t)\end{pmatrix}$$

are solutions of the considered β-differential system. Therefore,

$$x(t) = c_1 e_{1,\beta}(t) \begin{pmatrix} \cos_{\frac{1}{1+(\beta(t)-t)},\beta}(t) \\ \\ -\sin_{\frac{1}{1+(\beta(t)-t)},\beta}(t) \end{pmatrix} + c_2 e_{1,\beta}(t) \begin{pmatrix} \sin_{\frac{1}{1+(\beta(t)-t)},\beta}(t) \\ \\ \cos_{\frac{1}{1+(\beta(t)-t)},\beta}(t) \end{pmatrix},$$

where $c_1, c_2 \in \mathbb{R}$, is a general solution of the considered β-differential system.

Example 2.13 Consider the β-differential system

$$D_\beta x_1(t) = x_2(t)$$

$$D_\beta x_2(t) = x_3(t)$$

$$D_\beta x_3(t) = 2x_1(t) - 4x_2(t) + 3x_3(t), \quad t \in I.$$

Here

$$A = \begin{pmatrix} 0 & 1 & 0 \\ \\ 0 & 0 & 1 \\ \\ 2 & -4 & 3 \end{pmatrix}.$$

Then

$$\det(A - \lambda \mathscr{I}) = \det \begin{pmatrix} -\lambda & 1 & 0 \\ \\ 0 & -\lambda & 1 \\ \\ 2 & -4 & 3-\lambda \end{pmatrix}$$

$$= -\lambda^2(\lambda - 3) + 2 - 4\lambda$$

$$= -\left(\lambda^3 - 3\lambda^2 + 4\lambda - 2\right)$$

$$= -(\lambda - 1)(\lambda^2 - 2\lambda + 2),$$

whereupon

$$\det(A - \lambda \mathscr{I}) = 0$$

yields

$$\lambda_1 = 1,$$

$$\lambda_{2,3} = 1 \pm i.$$

Note that

$$\xi_1 = \begin{pmatrix} 1 \\ 1 \\ 1 \end{pmatrix} \quad \text{and}$$

$$\xi_2 = \begin{pmatrix} 1 \\ 1+i \\ 2i \end{pmatrix}$$

are eigenvectors corresponding to the eigenvalues $\lambda_1 = 1$ and $\lambda_2 = 1+i$, respectively. Note that

$$e_{1+i,\beta}(t) \begin{pmatrix} 1 \\ 1+i \\ 2i \end{pmatrix} = e_{1,\beta}(t) \left(\cos_{\frac{1}{1+(\beta(t)-t)},\beta}(t) + i \sin_{\frac{1}{1+(\beta(t)-t)},\beta}(t) \right) \begin{pmatrix} 1 \\ 1+i \\ 2i \end{pmatrix}$$

$$= e_{1,\beta}(t) \left(\begin{pmatrix} \cos_{\frac{1}{1+(\beta(t)-t)},\beta}(t) \\ (1+i)\cos_{\frac{1}{1+(\beta(t)-t)},\beta}(t) \\ 2i\cos_{\frac{1}{1+(\beta(t)-t)},\beta}(t) \end{pmatrix} + i \begin{pmatrix} \sin_{\frac{1}{1+(\beta(t)-t)},\beta}(t) \\ (1+i)\sin_{\frac{1}{1+(\beta(t)-t)},\beta}(t) \\ 2i\sin_{\frac{1}{1+(\beta(t)-t)},\beta}(t) \end{pmatrix} \right)$$

$$= e_{1,\beta}(t) \left(\begin{pmatrix} \cos_{\frac{1}{1+(\beta(t)-t)},\beta}(t) \\ (1+i)\cos_{\frac{1}{1+(\beta(t)-t)},\beta}(t) \\ 2i\cos_{\frac{1}{1+(\beta(t)-t)},\beta}(t) \end{pmatrix} + \begin{pmatrix} i\sin_{\frac{1}{1+(\beta(t)-t)},\beta}(t) \\ (-1+i)\sin_{\frac{1}{1+(\beta(t)-t)},\beta}(t) \\ -2\sin_{\frac{1}{1+(\beta(t)-t)},\beta}(t) \end{pmatrix} \right)$$

$$= e_{1,\beta}(t) \left(\begin{pmatrix} \cos_{\frac{1}{1+(\beta(t)-t)},\beta}(t) \\ \cos_{\frac{1}{1+(\beta(t)-t)},\beta}(t) - \sin_{\frac{1}{1+(\beta(t)-t)},\beta}(t) \\ -2\sin_{\frac{1}{1+(\beta(t)-t)},\beta}(t) \end{pmatrix} \right.$$

$$\left. + i \begin{pmatrix} \sin_{\frac{1}{1+(\beta(t)-t)},\beta}(t) \\ \cos_{\frac{1}{1+(\beta(t)-t)},\beta}(t) + \sin_{\frac{1}{1+(\beta(t)-t)},\beta}(t) \\ 2\cos_{\frac{1}{1+(\beta(t)-t)},\beta}(t) \end{pmatrix} \right).$$

Consequently,

$$x(t) = e_{1,\beta}(t)\left(c_1\begin{pmatrix}1\\1\\1\end{pmatrix}+c_2\begin{pmatrix}\cos_{\frac{1}{1+(\beta(t)-t)},\beta}(t)\\ \cos_{\frac{1}{1+(\beta(t)-t)},\beta}(t)-\sin_{\frac{1}{1+(\beta(t)-t)},\beta}(t)\\ -2\sin_{\frac{1}{1+(\beta(t)-t)},\beta}(t)\end{pmatrix}\right.$$

$$\left.+c_3\begin{pmatrix}\sin_{\frac{1}{1+(\beta(t)-t)},\beta}(t)\\ \cos_{\frac{1}{1+(\beta(t)-t)},\beta}(t)+\sin_{\frac{1}{1+(\beta(t)-t)},\beta}(t)\\ 2\cos_{\frac{1}{1+(\beta(t)-t)},\beta}(t)\end{pmatrix}\right),$$

where $c_1, c_2, c_3 \in \mathbb{R}$, is a general solution of the considered β-differential system.

Exercise 2.8 Find a general solution of the β-differential system

$$D_\beta x_1(t) = x_1(t) - 2x_2(t) + x_3(t)$$

$$D_\beta x_2(t) = -x_1(t) + x_3(t)$$

$$D_\beta x_3(t) = x_1(t) - 2x_2(t) + x_3(t), \quad t \in I.$$

Theorem 2.16 (The β-Putzer Algorithm)

Let $A \in \mathscr{R}_\beta$ be a constant $n \times n$-matrix. If $\lambda_1, \lambda_2, \ldots, \lambda_n$ are the eigenvalues of A. Then

$$e_{A,\beta}(t) = \sum_{k=0}^{n-1} r_{k+1}(t)P_k,$$

where

$$r(t) = \begin{pmatrix}r_1(t)\\ \vdots\\ r_n(t)\end{pmatrix}$$

is the solution of the IVP

$$D_\beta r = \begin{pmatrix}\lambda_1 & 0 & 0 & \ldots & 0\\ 1 & \lambda_2 & 0 & \ldots & 0\\ 0 & 1 & \lambda_3 & \ldots & 0\\ \vdots & \vdots & \vdots & \vdots & \vdots\\ 0 & 0 & 0 & \ldots & \lambda_n\end{pmatrix} r, \quad r(t_0) = \begin{pmatrix}1\\0\\0\\ \vdots\\ 0\end{pmatrix}, \tag{2.4}$$

and the P-matrices are recursively defined by

$$P_0 = I \quad \textit{and}$$

$$P_{k+1} = (A - \lambda_{k+1}\mathscr{I})P_k, \quad 0 \le k \le n-1.$$

Proof Since A is β-regressive, we have that all eigenvalues of A are β-regressive. Therefore, the IVP (2.4) has a unique solution. We set

$$X(t) = \sum_{k=0}^{n-1} r_{k+1}(t)P_k. \tag{2.5}$$

We have

$$\begin{aligned}
P_1 &= (A - \lambda_1\mathscr{I})P_0 \\
&= (A - \lambda_1\mathscr{I}), \\
P_2 &= (A - \lambda_2\mathscr{I})P_1 \\
&= (A - \lambda_2 I)(A - \lambda_1\mathscr{I}), \\
&\vdots \\
P_n &= (A - \lambda_n\mathscr{I})P_{n-1} \\
&= (A - \lambda_n\mathscr{I})\ldots(A - \lambda_1\mathscr{I}) \\
&= 0.
\end{aligned}$$

Therefore,

$$D_\beta X(t) = \sum_{k=0}^{n-1} D_\beta r_{k+1}(t)P_k$$

and

$$\begin{aligned}
D_\beta X(t) - AX(t) &= \sum_{k=0}^{n-1} D_\beta r_{k+1}(t)P_k - A\sum_{k=0}^{n-1} r_{k+1}(t)P_k \\
&= D_\beta r_1(t)P_0 + \sum_{k=1}^{n-1} D_\beta r_{k+1}(t)P_k \\
&\quad -A\sum_{k=0}^{n-1} r_{k+1}(t)P_k \\
&= \lambda_1 r_1(t)P_0 + \sum_{k=1}^{n-1} (r_k(t) + \lambda_{k+1} r_{k+1}(t))P_k \\
&\quad -\sum_{k=1}^{n-1} r_{k+1}(t)AP_k \\
&= \sum_{k=1}^{n-1} r_k(t)P_k + \lambda_1 r_1(t)P_0 \\
&\quad +\sum_{k=1}^{n-1} \lambda_{k+1} r_{k+1}(t)P_k - \sum_{k=0}^{n-1} r_{k+1}(t)AP_k \\
&= \sum_{k=1}^{n-1} r_k(t)P_k + \sum_{k=0}^{n-1} \lambda_{k+1} r_{k+1}(t)P_k \\
&\quad -\sum_{k=0}^{n-1} r_{k+1}(t)AP_k \\
&= \sum_{k=1}^{n-1} r_k(t)P_k - \sum_{k=0}^{n-1} (A - \lambda_{k+1} I) r_{k+1}(t)P_k \\
&= \sum_{k=1}^{n-1} r_k(t)P_k - \sum_{k=0}^{n-1} r_{k+1}(t)P_{k+1} \\
&= -r_n(t)P_n \\
&= 0, \quad t \in I.
\end{aligned}$$

Thus,

$$D_\beta X(t) = AX(t), \quad t \in I.$$

Also,

$$
\begin{aligned}
X(t_0) &= \sum_{k=0}^{n-1} r_{k+1}(t_0)P_k \\
&= r_1(t_0)P_0 \\
&= \mathscr{I}.
\end{aligned}
$$

Hence, X defined in (2.5) is really a solution of the IVP (2.4). This completes the proof. □

Example 2.14 Consider the β-differential system

$$
\begin{aligned}
D_\beta x_1(t) &= 2x_1(t) + x_2(t) + 2x_3(t) \\
D_\beta x_2(t) &= 4x_1(t) + 2x_2(t) + 4x_3(t) \\
D_\beta x_3(t) &= 2x_1(t) + x_2(t) + 2x_3(t), \quad t \in I.
\end{aligned}
$$

Here

$$
A = \begin{pmatrix} 2 & 1 & 2 \\ 4 & 2 & 4 \\ 2 & 1 & 2 \end{pmatrix}.
$$

Then

$$
\begin{aligned}
\det(A - \lambda \mathscr{I}) &= \det \begin{pmatrix} 2-\lambda & 1 & 2 \\ 4 & 2-\lambda & 4 \\ 2 & 1 & 2-\lambda \end{pmatrix} \\
&= -(\lambda-2)^3 + 8 + 8 + 4(\lambda-2) + 4(\lambda-2) + 4(\lambda-2) \\
&= -(\lambda-2)^3 + 12(\lambda-2) + 16 \\
&= -\left(\lambda^3 - 6\lambda^2 + 12\lambda - 8 - 12\lambda + 24 - 16\right) \\
&= -\left(\lambda^3 - 6\lambda^2\right) \\
&= -\lambda^2(\lambda-6),
\end{aligned}
$$

whereupon

$$\det(A - \lambda \mathscr{I}) = 0$$

yields

$$\lambda_1 = 0,$$

$$\lambda_2 = 0, \quad \text{and}$$

$$\lambda_3 = 6.$$

Consider the following IVPs:

$$D_\beta r_1(t) = 0, \quad r_1(t_0) = 1,$$

$$D_\beta r_2(t) = r_1(t), \quad r_2(t_0) = 0,$$

$$D_\beta r_3(t) = r_2(t) + 6r_3(t), \quad r_3(t_0) = 0.$$

Then we have

$$r_1(t) = 1, \quad t \in I,$$

$$D_\beta r_2(t) = 1, \quad r_2(t_0) = 0, \quad t \in I.$$

Thus,

$$r_2(t) = t - t_0, \quad t \in I,$$

and

$$D_\beta r_3(t) = t - t_0 + 6r_3(t), \quad r_3(t_0) = 0, \quad t \in I.$$

This gives

$$r_3(t) = \int_{s_0}^{t} e_{6,\beta}(\beta(\tau))(\tau - t_0) d_\beta \tau, \quad t \in I.$$

Next, we have

$$P_0 = \begin{pmatrix} 1 & 0 & 0 \\ 0 & 1 & 0 \\ 0 & 0 & 1 \end{pmatrix},$$

$$\begin{aligned}
P_1 &= (A - \lambda_1 \mathscr{I})P_0 \\
&= AP_0 \\
&= A \\
&= \begin{pmatrix} 2 & 1 & 2 \\ 4 & 2 & 4 \\ 2 & 1 & 2 \end{pmatrix}, \\
P_2 &= (A - \lambda_1 I)(A - \lambda_2 I) \\
&= A^2 I \\
&= A^2 \\
&= \begin{pmatrix} 2 & 1 & 2 \\ 4 & 2 & 4 \\ 2 & 1 & 2 \end{pmatrix} \begin{pmatrix} 2 & 1 & 2 \\ 4 & 2 & 4 \\ 2 & 1 & 2 \end{pmatrix} \\
&= \begin{pmatrix} 12 & 6 & 12 \\ 24 & 12 & 24 \\ 12 & 6 & 12 \end{pmatrix}, \quad \text{and} \\
P_3 &= 0.
\end{aligned}$$

Hence

$$\begin{aligned}
e_{A,\beta}(t) &= r_1(t)P_0 + r_2(t)P_1 + r_3(t)P_2 \\
&= \begin{pmatrix} 1 & 0 & 0 \\ 0 & 1 & 0 \\ 0 & 0 & 1 \end{pmatrix} + (t - t_0) \begin{pmatrix} 2 & 1 & 2 \\ 4 & 2 & 4 \\ 2 & 1 & 2 \end{pmatrix} \\
&\quad + \left(\int_{t_0}^{t} e_{6,\beta}(\beta(\tau))(\tau - t_0) d_\beta \tau \right) \begin{pmatrix} 12 & 6 & 12 \\ 24 & 12 & 24 \\ 12 & 6 & 12 \end{pmatrix},
\end{aligned}$$

and

$$\begin{pmatrix} x_1(t) \\ x_2(t) \\ x_3(t) \end{pmatrix} = e_{A,\beta}(t) \begin{pmatrix} c_1 \\ c_2 \\ c_3 \end{pmatrix},$$

where $c_1, c_2, c_3 \in \mathbb{R}$, is a general solution of the considered β-differential system.

Exercise 2.9 Using the β-Putzer algorithm, find $e_{A,\beta}(t)$, where

1.

$$A = \begin{pmatrix} 1 & 2 \\ -1 & 3 \end{pmatrix},$$

2.

$$A = \begin{pmatrix} 1 & -1 & 1 \\ 1 & 0 & 2 \\ -1 & 1 & 1 \end{pmatrix}.$$

2.5 Nonlinear Systems

In this section, we will investigate the following nonlinear system:

$$D_\beta y_j = f_j(t,y), \quad j \in \{1,\ldots,n\}, \quad t \in [a,b], \tag{2.6}$$

subject to the initial condition

$$y_j(a) = y_{0j}, \quad j \in \{1,\ldots,n\}, \tag{2.7}$$

where

(A1) $[a,b] \subset I$, $s_0 \in [a,b]$,
(A2) $f = (f_1,\ldots,f_n)$, $f_j \in \mathscr{C}([a,b] \times \mathbb{R})$, $j \in \{1,\ldots,n\}$, $y_0 = (y_{01},\ldots,y_{0n}) \in \mathbb{R}^n$, $y = (y_1,\ldots,y_n)$ is the unknown function.

Denote $X = (\mathscr{C}([a,b]))^n$. In $\mathscr{C}([a,b] \times \mathbb{R})$, we define a norm

$$\|u\|_1 = \sup_{(t,z)\in[a,b]\times\mathbb{R}} |u(t,z)|,$$

provided it exists. In X, we define a norm

$$\|y\| = \max_{j\in\{1,\ldots,n\}} \|y_j\|_1.$$

In this section, we will start our investigations with the following useful lemma.

Lemma 2.1 *Assume that* (A1) *and* (A2) *hold. If* $y \in (\mathscr{C}_\beta([a,b]))^n$ *is a solution to the IVP* (2.6), (2.7), *then it satisfies the integral equation*

$$y_j(t) = y_{0j} + \int_a^t f_j(s, y(s)) d_\beta s, \quad t \in [a,b], \quad j \in \{1, \ldots, n\}. \tag{2.8}$$

Proof We β-integrate equation (2.6), and using the initial condition (2.7), we obtain the integral equation (2.8). This completes the proof. □

Lemma 2.2 *Assume that* (A1) *and* (A2) *hold. If* $y \in (\mathscr{C}_\beta([a,b]))^n$ *satisfies the integral equation* (2.8), *then it is a solution to the IVP* (2.6), (2.7).

Proof Since $f_j \in \mathscr{C}([a,b] \times \mathbb{R})$, $j \in \{1, \ldots, n\}$, we have that

$$\int_a^t f_j(s, y(s)) d_\beta s \in \mathscr{C}_\beta^1([a,b]), \quad j \in \{1, \ldots, n\}.$$

Then $y_j \in \mathscr{C}_\beta^1([a,b])$, $j \in \{1, \ldots, n\}$. Now, we β-differentiate the integral equation (2.8) and we get equation (2.6). We put $t = a$ into (2.8) and we find (2.7). Thus, y is a solution to the IVP (2.6), (2.7). This completes the proof. □

In addition, suppose the following:

(A3)

$$|f_j(t, y^1) - f_j(t, y^2)| \le A \sum_{l=1}^n |y_l^1 - y_l^2|, \quad t \in [a,b], \quad y^1, y^2 \in \mathbb{R}^n,$$

$j \in \{1, \ldots, n\}$, for some positive constant A.

(A4) $|f_j(t,y)| \le M$, $t \in [a,b]$, $y \in \mathbb{R}^n$, $j \in \{1, \ldots, n\}$, for some nonnegative constant M.

Theorem 2.17 *Assume that* (A1)-(A4) *hold. Then the IVP* (2.6), (2.7) *has a unique solution* $y \in (\mathscr{C}_\beta([a,b]))^n$.

Proof We define the sequence

$$y_{j0}(t) = y_{j0},$$

$$y_{jl}(t) = y_{j0}(t) + \int_a^t f(s, y_{l-1}(s)) d_\beta s, \quad t \in [a,b], \quad l \in \mathbb{N}.$$

We have

$$\begin{aligned}|y_{j1}(t)-y_{j0}(t)| &= \left|\int_a^t f_j(s,y_0(s))d_\beta s\right| \\ &\le \int_a^t |f_j(s,y_0(s))|d_\beta s \\ &\le M\int_a^t d_\beta s \\ &= Mh_1(t,a), \quad t\in[a,b], \quad j\in\{1,\ldots,n\}.\end{aligned}$$

Assume that

$$|y_{jl-1}(t)-y_{jl-2}(t)| \le Mn^{l-2}A^{l-2}h_{l-1}(t,a), \quad t\in[a,b], \quad j\in\{1,\ldots,n\}, \tag{2.9}$$

for some $l\in\mathbb{N}$, $l\ge 2$. We will prove that

$$|y_{jl}(t)-y_{jl-1}(t)| \le Mn^{l-1}A^{l-1}h_l(t,a), \quad t\in[a,b], \quad j\in\{1,\ldots,n\}.$$

Really, we have

$$\begin{aligned}|y_{jl}(t)-y_{jl-1}(t)| &= \left|\int_a^t (f_j(s,y_l(s))-f_j(s,y_{l-1}(s)))d_\beta s\right| \\ &\le \int_a^t \left|f_j(s,y_l(s))-f_j(s,y_{l-1}(s))\right|d_\beta s \\ &\le A\sum_{m=1}^n \int_a^t |y_{ml}(s)-y_{ml-1}(s)|d_\beta s \\ &\le AMn^{l-2}A^{l-2}\sum_{m=1}^n \int_a^t h_{l-1}(s,a)d_\beta s \\ &= MA^{l-1}n^{l-1}h_l(t,a), \quad t\in[a,b], \quad j\in\{1,\ldots,n\}.\end{aligned}$$

Thus, (2.9) holds for any $l \in \mathbb{N}, l \geq 2$. Note that

$$\left| \lim_{l \to \in fty} \left(y_{jl}(t) - y_{j0}(t) \right) \right| = \left| \sum_{l=1}^{\in fty} \left(y_{jl}(t) - y_{jl-1}(t) \right) \right|$$

$$\leq \sum_{l=1}^{\in fty} \left| y_{jl}(t) - y_{jl-1}(t) \right|$$

$$\leq M \sum_{l=1}^{\in fty} A^{l-1} n^{l-1} h_l(t,a)$$

$$= \frac{M}{An} \sum_{l=1}^{\in fty} A^l n^l h_l(t,a)$$

$$\leq \frac{M}{An} \sum_{l=1}^{\in fty} A^l n^l h_l(b,a)$$

$$= \frac{M}{An} e_{An,\beta}(b,a)$$

$$< \in fty, \quad j \in \{1, \ldots, n\}.$$

So, the series

$$\sum_{l=1}^{\in fty} \left(y_{jl}(t) - y_{jl-1}(t) \right), \quad j \in \{1, \ldots, n\},$$

is uniformly convergent on $[a,b]$. Hence, there exists

$$\lim_{l \to \in fty} (y_{jl}(t) - y_{j0}(t)) = \sum_{l=1}^{\in fty} (y_{jl}(t) - y_{jl-1}(t)), \quad j \in \{1, \ldots, n\},$$

$t \in [a,b]$, and then there exists

$$y(t) = \lim_{l \to \in fty} y_l(t), \quad t \in [a,b],$$

and $y_j(t)$, $t \in [a,b]$, $j \in \{1, \ldots, n\}$, satisfy (2.8). From here and Lemma 2.2, we conclude that $y(t)$, $t \in [a,b]$, is a solution to the IVP (2.6), (2.7). Assume that the IVP (2.6), (2.7) has another solution $z(t)$, $t \in [a,b]$. By Lemma 2.1, we have that

$$z_j(t) = y_{j0}(t) + \int_a^t f_j(s, z(s)) d_\beta s, \quad t \in [a,b], \quad j \in \{1, \ldots, n\}.$$

Note that

$$\begin{aligned}|z_j(t)-y_{j0}(t)| &= \left|\int_a^t f_j(s,z(s))d_\beta s\right|\\ &\le \int_a^t |f_j(s,z(s))|d_\beta s\\ &\le M\int_a^t d_\beta s\\ &= Mh_1(t,a),\quad t\in[a,b],\quad j\in\{1,\dots,n\}.\end{aligned}$$

Assume that

$$|y_{jl-1}(t)-z_j(t)|\le Mn^{l-2}A^{l-2}h_{l-1}(t,a),\quad t\in[a,b],\quad j\in\{1,\dots,n\},\tag{2.10}$$

for some $l\in\mathbb{N}$, $l\ge 2$. We will prove that

$$|y_{jl}(t)-z_j(t)|\le Mn^{l-1}A^{l-1}h_l(t,a),\quad t\in[a,b],\quad j\in\{1,\dots,n\}.$$

In fact, we have

$$\begin{aligned}|y_{jl}(t)-z_j(t)| &= \left|\int_a^t (f_j(s,y_{l-1}(s))-f_j(s,z(s)))d_\beta s\right|\\ &\le \int_a^t \left|f_j(s,z(s))-f_j(s,y_{l-1}(s))\right|d_\beta s\\ &\le A\sum_{m=1}^n\int_a^t |z_m(s)-y_{ml-1}(s)|d_\beta s\\ &\le AMn^{l-2}A^{l-2}\sum_{m=1}^n\int_a^t h_{l-1}(s,a)d_\beta s\\ &= MA^{l-1}n^{l-1}h_l(t,a),\quad t\in[a,b],\quad j\in\{1,\dots,n\}.\end{aligned}$$

Thus, (2.10) holds for any $l\in\mathbb{N}$, $l\ge 2$. Because

$$\lim_{l\to\in fty} MA^ln^lh_l(t,a)=0,\quad t\in[a,b],$$

we conclude that

$$\begin{aligned} y_j(t) - z_j(t) &= \lim_{l \to \in fty} (y_{jl}(t) - z_j(t)) \\ &= 0, \quad t \in [a,b], \quad j \in \{1, \ldots, n\}. \end{aligned}$$

This completes the proof. □

2.6 Advanced Practical Problems

Problem 2.1 Let $I = \mathbb{R}$ and define $\beta(t) = 3t$ and

$$A(t) = \begin{pmatrix} \frac{t^2+2}{t+1} & t^2+3t \\ 4t-1 & 3t \end{pmatrix}, \quad t \in I.$$

Then find $D_\beta A(t)$, $t \in I$.

Problem 2.2 Let $I = \mathbb{R}$ and define $\beta(t) = 3t$,

$$A(t) = \begin{pmatrix} t+10 & t^2-2t+2 \\ t & t^2+t+1 \end{pmatrix}, \quad \text{and}$$

$$B(t) = \begin{pmatrix} t-2 & t+1 \\ t^2 & t^3-1 \end{pmatrix}, \quad t \in I.$$

Then show that

$$D_\beta(AB)(t) = D_\beta A(t) B^\beta(t) + A(t) D_\beta B(t), \quad t \in I.$$

Problem 2.3 Let $I = \mathbb{R}$ and define $\beta(t) = \frac{1}{3}t + 7$, and

$$A(t) = \begin{pmatrix} t^2+2t+2 & t+1 \\ \frac{1}{t+1} & t^2+2 \end{pmatrix}, \quad t \in I.$$

Then show that

$$(A^\beta)^{-1}(t) = (A^{-1})^\beta(t), \quad t \in I.$$

Problem 2.4 Let $I = \mathbb{R}$ and define $\beta(t) = 3t$,

$$A(t) = \begin{pmatrix} 1 & -1 \\ 0 & 1 \end{pmatrix}, \quad \text{and}$$

$$B(t) = \begin{pmatrix} 2 & 0 \\ -1 & 1 \end{pmatrix}, \quad t \in I.$$

Then find

$$(A \oplus_\beta B)(t), \quad t \in I.$$

Problem 2.5 Find a general solution of the β-differential system

$$D_\beta x_1(t) = 2x_1(t) + 3x_2(t)$$

$$D_\beta x_2(t) = x_1(t) + 4x_2(t), \quad t \in I.$$

Problem 2.6 Find a general solution of the β-differential system

$$D_\beta x_1(t) = -x_1(t) - x_2(t) - x_3(t)$$

$$D_\beta x_2(t) = x_1(t) - x_2(t) + 3x_3(t)$$

$$D_\beta x_3(t) = x_1(t) - x_2(t) + 4x_3(t), \quad t \in I.$$

Problem 2.7 Using the β-Putzer algorithm, find $e_{A,\beta}(t)$, where

1.
$$A = \begin{pmatrix} 2 & 3 \\ 1 & -4 \end{pmatrix},$$

2.
$$A = \begin{pmatrix} -1 & 2 & 3 \\ 1 & 1 & -4 \\ 1 & -1 & 2 \end{pmatrix}.$$

Chapter 3
Functionals

Suppose that $I \subset \mathbb{R}$ and $\beta : I \to I$ is a general first kind or second kind, and $t_0 \in I$ is its fixed point. Let $a, b \in I$, $a < b$.

3.1 Definition for Functionals

Definition 3.1 (Functional) By a functional, we mean a correspondence which as signs a definite(real) number to each function belonging to some class.

Example 3.1 Let $y \in \mathscr{C}_\beta^1([a,b])$. The formula

$$J(y) = \int_a^b (D_\beta y(t))^2 \, d_\beta t$$

defines a functional on $\mathscr{C}_\beta^1([a,b])$.

Example 3.2 Let $y \in \mathscr{C}_\beta^1([a,b])$, $F \in \mathscr{C}([a,b] \times \mathbb{R} \times \mathbb{R})$. The expression

$$J(y) = \int_a^b F(t, y(t), D_\beta y(t)) \, d_\beta t$$

defines a functional on $\mathscr{C}_\beta^1([a,b])$.

Definition 3.2 Let X be a linear normed space with a norm $\|\cdot\|$.

1. The functional $J : X \to \mathbb{R}$ is said to be continuous at the point $y_1 \in X$ if for each $\varepsilon > 0$ there exists a $\delta = \delta(\varepsilon) > 0$ such that

$$|J(y) - J(y_1)| < \varepsilon$$

provided that $\|y - y_1\| < \delta$.
2. The functional $J : X \to \mathbb{R}$ is said to be continuous on X if it is continuous at each point of X.
3. The functional $J : X \to \mathbb{R}$ is said to be linear if

$$J(\alpha_1 y_1 + \alpha_2 y_2) = \alpha_1 J(y_1) + \alpha_2 J(y_2)$$

for any $\alpha_1, \alpha_2 \in \mathbb{R}$ and for any $y_1, y_2 \in X$.

Example 3.3 For $y \in \mathscr{C}_\beta([a,b])$, the expression

$$J(y) = \int_a^b y(t) d_\beta t$$

defines a linear functional on $\mathscr{C}_\beta([a,b])$.

3.2 Self-Adjoint Second Order Matrix Equations

Suppose that $R, P \in \mathscr{C}_\beta(I, \mathbb{R}^{n\times n})$ is such that $R(t)$ is symmetric and invertible, $P(t)$ is symmetric for each $t \in I$. Consider the second order matrix equation

$$D_\beta \left(R(t) D_\beta Y\right) + P(t) Y^\beta = 0. \tag{3.1}$$

Definition 3.3 We call $Y \in \mathscr{C}_\beta^1(I, \mathbb{R}^{n\times n})$ a solution of equation (3.1) provided

$$D_\beta \left((R D_\beta Y) + P Y^\beta \right)(t) = 0$$

holds for all $t \in I$.

Remark 3.1 If $Y \in \mathscr{C}_\beta(I, \mathbb{R}^{n\times n})$ is a solution of equation (3.1), then

$$R D_\beta Y \in \mathscr{C}_\beta^1(I, \mathbb{R}^{n\times n}).$$

Theorem 3.1 (The Wronskian Identity) *Let Y and Y_1 be solutions of* (3.1). *Then*

$$Y^T R D_\beta Y_1 - (D_\beta Y)^T R Y_1$$

is a constant.

Proof We have

$$D_\beta (R D_\beta Y) = -P Y^\beta,$$

$$D_\beta \left((D_\beta Y)^T R \right) = - \left(Y^\beta \right)^T P,$$

and

$$\begin{aligned} D_\beta \left(Y^T R D_\beta Y_1 - (D_\beta Y)^T R Y_1 \right) &= D_\beta \left(Y^T R D_\beta Y_1 \right) - D_\beta \left((D_\beta Y)^T R Y_1 \right) \\ &= \left(Y^\beta \right)^T D_\beta (R D_\beta Y_1) + (D_\beta Y)^T R D_\beta Y_1 \\ &\quad - D_\beta \left((D_\beta Y)^T R \right) Y_1^\beta - (D_\beta Y)^T R D_\beta Y_1 \\ &= \left(Y^\beta \right)^T \left(-P Y_1^\beta \right) + \left(Y^\beta \right)^T P Y_1^\beta \\ &= 0. \end{aligned}$$

This completes the proof. □

Now we set

$$Z = \begin{pmatrix} Y \\ R D_\beta Y \end{pmatrix} \quad on \quad I \quad and \quad S = \begin{pmatrix} 0 & R^{-1} \\ -P & -\mu P R^{-1} \end{pmatrix} \quad on \quad I.$$

Theorem 3.2 *Y solves* (3.1) *if and only if Z solves*

$$D_\beta Z = S(t) Z \quad on \quad I. \tag{3.2}$$

Proof 1. Let Y solves (3.1). Then

$$\begin{aligned} D_\beta Z(t) &= D_\beta \begin{pmatrix} Y \\ R D_\beta Y \end{pmatrix} (t) \\ &= \begin{pmatrix} D_\beta Y(t) \\ D_\beta (R D_\beta Y)(t) \end{pmatrix} \\ &= \begin{pmatrix} D_\beta Y(t) \\ -P(t) Y^\beta (t) \end{pmatrix}, \end{aligned}$$

$$
\begin{aligned}
S(t)Z(t) &= \begin{pmatrix} 0 & R^{-1}(t) \\ -P(t) & -(\beta(t)-t)P(t)R^{-1}(t) \end{pmatrix} \begin{pmatrix} Y(t) \\ R(t)D_\beta Y(t) \end{pmatrix} \\
&= \begin{pmatrix} D_\beta Y(t) \\ -P(t)Y(t) - (\beta(t)-t)P(t)D_\beta Y(t) \end{pmatrix} \\
&= \begin{pmatrix} D_\beta Y(t) \\ -P(t)\left(Y(t) + (\beta(t)-t)D_\beta Y(t)\right) \end{pmatrix} \\
&= \begin{pmatrix} D_\beta Y(t) \\ -P(t)Y^\beta(t) \end{pmatrix}, \quad t \in I.
\end{aligned}
$$

Therefore, Z solves (3.2).

2. Let Z solve (3.2). Then, using the computations in the previous point, we have

$$
\begin{pmatrix} D_\beta Y(t) \\ D_\beta \left(RD_\beta Y\right)(t) \end{pmatrix} = \begin{pmatrix} D_\beta Y(t) \\ -P(t)Y^\beta(t) \end{pmatrix}, \quad t \in I,
$$

whereupon

$$
D_\beta \left(RD_\beta Y\right)(t) = -P(t)Y^\beta(t), \quad t \in I,
$$

i.e., Y solves (3.1). This completes the proof. □

Theorem 3.3 *S is a regressive matrix on I.*

Proof We have

$$
\begin{aligned}
I + (\beta(t)-t)S(t) &= \begin{pmatrix} I & 0 \\ 0 & I \end{pmatrix} + (\beta(t)-t)\begin{pmatrix} 0 & R^{-1}(t) \\ -P(t) & -(\beta(t)-t)P(t)R^{-1}(t) \end{pmatrix} \\
&= \begin{pmatrix} I & (\beta(t)-t)R^{-1}(t) \\ -(\beta(t)-t)P(t) & I - ((\beta(t)-t))^2 P(t)R^{-1}(t) \end{pmatrix},
\end{aligned}
$$

$$
\begin{aligned}
det(I + (\beta(t)-t)S(t)) &= 1 - ((\beta(t)-t))^2 det(P(t)R^{-1}(t)) \\
&\quad + ((\beta(t)-t))^2 det(P(t)R^{-1}(t)) \\
&= 1, \quad t \in I.
\end{aligned}
$$

This completes the proof. □

Therefore, for any choice of a $2n \times 2n$ matrix Z_a, the problem

$$
D_\beta Z = S(t)Z, \quad Z(a) = Z_a,
$$

has a unique solution. From here, using Theorem 3.2, for any choice of $2n \times 2n$ matrices Y_a and Y_a', the problem

$$D_\beta\left(R(t)D_\beta Y\right) + P(t)Y^\beta = 0, \quad Y(a) = Y_a, \quad D_\beta Y(a) = Y_a',$$

has a unique solution Y. Let $\tilde{Y}$ be the unique solution of the initial value problem

$$D_\beta\left(R(t)D_\beta Y\right) + P(t)Y^\beta = 0, \quad Y(a) = 0, \quad D_\beta Y(a) = R^{-1}(a). \tag{3.3}$$

Definition 3.4 This $\tilde{Y}$ is called the principal solution of equation (3.1).

Theorem 3.4 *The matrices*

$$\tilde{Q}(t) = \left(RD_\beta\tilde{Y}\tilde{Y}^{-1}\right)(t) \quad \textit{and} \quad \tilde{D}(t) = \left(\tilde{Y}\left(\tilde{Y}^\beta\right)^{-1}R^{-1}\right)(t), \quad t \in I,$$

are symmetric matrices.

Proof By Theorem 3.1, the Wronskian identity, we have

$$\begin{aligned}
\left(\tilde{Y}^T RD_\beta\tilde{Y} - \left(D_\beta\tilde{Y}\right)^T R\tilde{Y}\right)(t) &= \left(\tilde{Y}^T RD_\beta\tilde{Y} - \left(D_\beta\tilde{Y}\right)^T R\tilde{Y}\right)(a)\\
&= \left(\tilde{Y}^T RD_\beta\tilde{Y}\right)(a)\\
&= \left(\tilde{Y}^T RR^{-1}\right)(a)\\
&= \tilde{Y}^T(a)\\
&= 0, \quad t \in I.
\end{aligned}$$

Therefore,

$$\left(\tilde{Y}^T RD_\beta\tilde{Y}\right)(t) = \left(\left(D_\beta\tilde{Y}\right)^T R\tilde{Y}\right)(t), \quad t \in I,$$

i.e., the matrix $\tilde{Y}^T RD_\beta\tilde{Y}$ is a symmetric matrix. Hence,

$$\left(\tilde{Y}^T RD_\beta\tilde{Y}\tilde{Y}^{-1}\right)(t) = \left(\left(D_\beta\tilde{Y}\right)^T R\right)(t),$$

and

$$\begin{aligned}
\left(RD_\beta\tilde{Y}\tilde{Y}^{-1}\right)(t) &= \left(\left(\tilde{Y}^T\right)^{-1}\left(D_\beta\tilde{Y}\right)^T R\right)(t)\\
&= \left(RD_\beta\tilde{Y}\tilde{Y}^{-1}\right)^T(t).
\end{aligned}$$

Consequently $\tilde{Q}(t), t \in I$, is a symmetric matrix. Note that

$$\left(R\tilde{Y}^{\beta}\tilde{Y}\right)^{-1}(t) = \left(R\left(\tilde{Y} + (\beta - I)D_{\beta}\tilde{Y}\right)\tilde{Y}^{-1}\right)(t)$$

$$= R(t) + (\beta(t) - t)\left(RD_{\beta}\tilde{Y}\tilde{Y}^{-1}\right)(t)$$

$$= R(t) + (\beta(t) - t)\tilde{Q}(t), \quad t \in I.$$

From here, since R and $\tilde{Q}$ are symmetric matrices on I, we conclude that $RD_{\beta}\tilde{Y}\tilde{Y}^{-1}$ and $\tilde{D}$ are symmetric matrices on I. This completes the proof. □

Definition 3.5 Let Y be a solution of equation (3.1). We say that Y is a conjoined solution of (3.1) provided $Y^T RD_{\beta}Y$ is symmetric.

Definition 3.6 Two conjoined solutions Y and Y_1 of (3.1) are called normalized if

$$Y^T RD_{\beta}Y_1 - (D_{\beta}Y)^T RY_1 = I \quad on \quad I.$$

Definition 3.7 The solution $\tilde{Y}_1$ of the problem

$$D_{\beta}\left(R(t)D_{\beta}Y\right) + P(t)Y^{\beta} = 0, \quad Y(a) = -I, \quad D_{\beta}Y(a) = 0$$

is called associated solution of (3.1) at a.

Theorem 3.5 *$\tilde{Y}$ and $\tilde{Y}_1$ are normalized solutions.*

Proof By Theorem 3.1, the Wronskian identity, we have

$$\left(\left(\tilde{Y}_1^T RD_{\beta}\tilde{Y}_1\right)(a) - \left(\left(D_{\beta}\tilde{Y}_1\right)^T R\tilde{Y}_1\right)(a)\right)^T$$

$$= \left(-\left(\left(D_{\beta}\tilde{Y}_1\right)^T R\tilde{Y}_1\right)(a)\right)^T$$

$$= \left(\left(\left(D_{\beta}\tilde{Y}_1\right)^T R\right)(a)\right)^T$$

$$= RD_{\beta}\tilde{Y}_1(a)$$

$$= 0,$$

$$\left(\tilde{Y}_1^T RD_\beta \tilde{Y}_1\right)(t) = \left(\left(D_\beta \tilde{Y}_1\right)^T R\tilde{Y}_1\right)(t), \quad t \in I,$$

$$\begin{aligned}\left(\left(\tilde{Y}^T RD_\beta \tilde{Y}\right)(a) - \left(\left(D_\beta \tilde{Y}\right)^T R\tilde{Y}\right)(a)\right)^T &\\ &= \left(\tilde{Y}^T(a)R(a)D_\beta \tilde{Y}(a)\right)^T \\ &= \left(D_\beta \tilde{Y}\right)^T(a)R(a)\tilde{Y}(a) \\ &= 0,\end{aligned}$$

$$\left(\tilde{Y}^T RD_\beta \tilde{Y}\right)(a) = \left(\left(D_\beta \tilde{Y}\right)^T R\tilde{Y}\right)(t), \quad t \in I,$$

$$\begin{aligned}\left(\tilde{Y}RD_\beta \tilde{Y}_1 - \left(D_\beta \tilde{Y}\right)^T R\tilde{Y}_1\right)(t) &= \left(\tilde{Y}RD_\beta \tilde{Y}_1 - \left(D_\beta \tilde{Y}\right)^T R\tilde{Y}_1\right)(a) \\ &= -\left(\left(D_\beta \tilde{Y}\right)^T R\tilde{Y}_1\right)(a) \\ &= \left(\left(D_\beta \tilde{Y}\right)^T R\right)(a) \\ &= I, \quad t \in I.\end{aligned}$$

This completes the proof. □

Now we consider the matrix Riccati equation

$$Q^{D_\beta} + P(t) + Q^T\left(R(t) + (\beta(t) - t)Q\right)^{-1} Q = 0. \tag{3.4}$$

Theorem 3.6 *If* (3.1) *has a conjoined solution Y such that $Y(t)$ is invertible for all $t \in [a,b]$, then Q defined by*

$$Q(t) = R(t)D_\beta Y(t)Y^{-1}(t), \quad t \in [a,b], \tag{3.5}$$

is a symmetric solution of (3.4). *Conversely, if* (3.4) *has a symmetric solution on $[a,b]$, then there exists a conjoined solution Y of* (3.1) *such that $Y(t)$ is invertible for all $t \in [a,b]$ and* (3.5) *holds.*

Proof 1. Let Y be a conjoined solution of equation (3.1) such that $Y(t)$ is invertible for all $t \in [a,b]$. Let also, Q be defined by (3.5). By Theorem 3.4, it follows that $Q(t)$ is symmetric for all $t \in [a,b]^{\kappa}$. Also,

$$\begin{aligned}
D_\beta Q(t) &= D_\beta \left(RD_\beta YY^{-1}\right)(t) \\
&= D_\beta \left(RD_\beta Y\right)(t)\left(Y^\beta\right)^{-1}(t) + \left(RD_\beta Y\right)(t) D_\beta \left(Y^{-1}\right)(t) \\
&= D_\beta \left(RD_\beta Y\right)(t)\left(Y^\beta\right)^{-1}(t) - \left(RD_\beta Y\right)(t)\left(Y^\beta\right)^{-1}(t) D_\beta Y(t) Y^{-1}(t) \\
&= -P(t)Y^\beta(t)\left(Y^\beta\right)^{-1}(t) - \left(RD_\beta Y\right)(t)\left(Y^\beta\right)^{-1}(t) D_\beta Y(t) Y^{-1}(t) \\
&= -P(t) - \left(RD_\beta Y\right)(t)\left(Y^\beta\right)^{-1}(t) D_\beta Y(t) Y^{-1}(t), \quad t \in [a,b]
\end{aligned}$$

and

$$\begin{aligned}
&D_\beta Q(t) + P(t) + Q^T(t)\left(R(t) + (\beta(t)-t)Q(t)\right)^{-1} Q(t) \\
&= -P(t) - R(t)D_\beta Y(t)\left(Y^\beta\right)^{-1}(t) D_\beta Y(t) Y^{-1}(t) \\
&\quad + P(t) + R(t) D_\beta Y(t) Y^{-1}(t) \\
&\quad \times \left(R(t) + (\beta(t)-t) R(t) D_\beta Y(t) Y^{-1}(t)\right)^{-1} R(t) D_\beta Y(t) Y^{-1}(t) \\
&= -R(t)D_\beta Y(t)\left(Y^\beta\right)^{-1}(t) D_\beta Y(t) Y^{-1}(t) \\
&\quad + R(t) D_\beta Y(t) Y^{-1}(t)\left(I + (\beta(t)-t) D_\beta Y(t) Y^{-1}(t)\right)^{-1} \\
&\quad \times R^{-1}(t) R(t) D_\beta Y(t) Y^{-1}(t) \\
&= -R(t)D_\beta Y(t)\left(Y^\beta\right)^{-1}(t) D_\beta Y(t) Y^{-1}(t) \\
&\quad + R(t) D_\beta Y(t) Y^{-1}(t)\left(I + \left(Y^\beta(t) - Y(t)\right) Y^{-1}(t)\right)^{-1} D_\beta Y(t) Y^{-1}(t) \\
&= -R(t)D_\beta Y(t)\left(Y^\beta\right)^{-1}(t) D_\beta Y(t) Y^{-1}(t) \\
&\quad + R(t) D_\beta Y(t) Y^{-1}(t)\left(Y^\beta(t) Y^{-1}(t)\right)^{-1} D_\beta Y(t) Y^{-1}(t)
\end{aligned}$$

$$= -R(t)D_\beta Y(t)\left(Y^\beta\right)^{-1}(t)D_\beta Y(t)Y^{-1}(t)$$

$$+R(t)D_\beta Y(t)Y^{-1}(t)Y(t)\left(Y^\beta\right)^{-1}(t)D_\beta Y(t)Y^{-1}(t)$$

$$= -R(t)D_\beta Y(t)\left(Y^\beta\right)^{-1}(t)D_\beta Y(t)Y^{-1}(t)$$

$$+R(t)D_\beta Y(t)\left(Y^\beta\right)^{-1}(t)D_\beta Y(t)Y^{-1}(t)$$

$$= 0, \quad t \in [a,b].$$

2. Let $Q(t)$, $t \in [a,b]$, be a symmetric solution of equation (3.4). Let Y be the solution of the IVP

$$D_\beta Y = R^{-1}(t)Q(t)Y, \quad Y(t_0) = I. \tag{3.6}$$

Then Y is a fundamental solution of the equation

$$D_\beta Y = R^{-1}(t)Q(t)Y$$

and hence it is invertible on $[a,b]$. Also,

$$D_\beta\left(RD_\beta Y\right)(t) = D_\beta(QY)(t)$$

$$= D_\beta Q(t)Y^\beta(t) + Q(t)D_\beta Y(t)$$

$$= \left(-P(t) - Q(t)\left(R(t) + (\beta(t)-t)Q(t)\right)^{-1}Q(t)\right)Y^\beta(t)$$

$$+Q(t)R^{-1}(t)Q(t)Y(t)$$

$$= -P(t)Y^\beta(t) - Q(t)\left(R(t) + (\beta(t)-t)Q(t)\right)^{-1}Q(t)Y^\beta(t)$$

$$+Q(t)R^{-1}(t)Q(t)Y(t)$$

$$= -P(t)Y^\beta(t) - Q(t)\left(R(t) + (\beta(t)-t)Q(t)\right)^{-1}Q(t)Y^\beta(t)$$

$$+Q(t)\left(R(t) + (\beta(t)-t)Q(t)\right)^{-1}$$

$$\times\left(R(t) + (\beta(t)-t)Q(t)\right)R^{-1}(t)Q(t)Y(t)$$

$$= -P(t)Y^{\beta}(t) + Q(t)\left(R(t) + (\beta(t) - t)Q(t)\right)^{-1}$$

$$\times \left((R(t) + (\beta(t) - t)Q(t))\, R^{-1}(t)Q(t)Y(t) - Q(t)Y^{\beta}(t) \right)$$

$$= -P(t)Y^{\beta}(t) + Q(t)\left(R(t) + (\beta(t) - t)Q(t)\right)^{-1}$$

$$\times \left(Q(t)Y(t) + (\beta(t) - t)Q(t)R^{-1}(t)Q(t)Y(t) - Q(t)Y^{\beta}(t) \right)$$

$$= -P(t)Y^{\beta}(t), \quad t \in [a,b].$$

In the last equation we have used that

$$Y^{\beta}(t) - Y(t) = (\beta(t) - t)R^{-1}(t)Q(t)Y(t), \quad t \in [a,b].$$

Since

$$\left(Y^T R D_{\beta} Y\right)(t_0) = Q(t_0),$$

we extend Y to a conjoined solution of (3.1). This completes the proof. □

Theorem 3.7 (The Picone Identity)
Let $\alpha \in \mathbb{R}^n$ and Y and Y_1 be normalized joined solutions of equation (3.1) *such that Y is invertible on $[a,b]$. We put*

$$Q = RD_{\beta}YY^{-1} \quad on \quad [a,b] \quad and \quad D = Y\left(Y^{\beta}\right)^{-1} R^{-1} \quad on \quad [a,b].$$

Let $t \in [a,b]^{\kappa}$ and $y : [a,b] \to \mathbb{R}^n$ be differentiable at t. Then we have at t

$$D_{\beta}\left(y^T Q y + 2\alpha^T Y^{-1} y - \alpha^T Y^{-1} Y_1 \alpha\right) = \left(D_{\beta} y\right)^T R D_{\beta} y - \left(y^{\beta}\right)^T P y^{\beta}$$

$$- \left(RD_{\beta}y - Qy - \left(Y^{-1}\right)^T \alpha\right)^T D \left(RD_{\beta}y - Qy - \left(Y^{-1}\right)^T \alpha\right).$$

Proof We have at t

$$D_{\beta}YY^{-1} = R^{-1}Q,$$

$$QY = RD_{\beta}Y,$$

$$Y^T Q = \left(D_{\beta}Y\right)^T R,$$

$$
\begin{aligned}
D_\beta Q &= D_\beta\left(RD_\beta YY^{-1}\right) \\
&= D_\beta\left(RD_\beta Y\right)\left(Y^\beta\right)^{-1} + RD_\beta Y D_\beta\left(Y^{-1}\right) \\
&= -PY^\beta\left(Y^\beta\right)^{-1} - RD_\beta Y\left(Y^\beta\right)^{-1} D_\beta YY^{-1} \\
&= -P - RD_\beta Y\left(Y^\beta\right)^{-1} D_\beta YY^{-1} \\
&= -P - QY\left(Y^\beta\right)^{-1} D_\beta YY^{-1} \\
&= -P - QY\left(Y^\beta\right)^{-1} R^{-1}RD_\beta YY^{-1} \\
&= -P - QY\left(Y^\beta\right)^{-1} R^{-1}Q, \\
R^{-1}Q &= D_\beta YY^{-1},
\end{aligned}
$$

and

$$
\begin{aligned}
D_\beta\left(Y^{-1}Y_1\right) &= -\left(Y^\beta\right)^{-1} D_\beta YY^{-1}Y_1 + \left(Y^\beta\right)^{-1} D_\beta Y_1 \\
&= -\left(Y^\beta\right)^{-1} R^{-1}QY_1 + \left(Y^\beta\right)^{-1} R^{-1}RD_\beta Y_1 \\
&= -\left(Y^\beta\right)^{-1} R^{-1}\left(Y^{-1}\right)^T Y^T QY_1 \\
&\quad + \left(Y^\beta\right)^{-1} R^{-1}\left(Y^{-1}\right)^T Y^T RD_\beta Y_1 \\
&= \left(Y^\beta\right)^{-1} R^{-1}\left(Y^{-1}\right)^T\left(Y^T RD_\beta Y_1 - Y^T QY_1\right) \\
&= \left(Y^\beta\right)^{-1} R^{-1}\left(Y^{-1}\right)^T\left(Y^T RD_\beta Y_1 - \left(D_\beta Y\right)^T RY_1\right) \\
&= \left(Y^\beta\right)^{-1} R^{-1}\left(Y^{-1}\right)^T,
\end{aligned}
$$

and

$$D_\beta\left(y^TQy\right)+\left(RD_\beta y-Qy\right)^TY\left(Y^\beta\right)^{-1}R^{-1}\left(RD_\beta y-Qy\right)$$

$$=\left(D_\beta y\right)^TQy+\left(y^\beta\right)^TD_\beta Qy^\beta+\left(y^\beta\right)^TQD_\beta y$$

$$+\left(D_\beta y\right)^TRY\left(Y^\beta\right)^{-1}D_\beta y-\left(D_\beta y\right)^TRY\left(Y^\beta\right)^{-1}R^{-1}Qy$$

$$-y^TQY\left(Y^\beta\right)^{-1}D_\beta y+y^TQY\left(Y^\beta\right)^{-1}R^{-1}Qy$$

$$=\left(D_\beta y\right)^TQy-\left(y^\beta\right)^TPy^\beta-\left(y^\beta\right)^TQY\left(Y^\beta\right)^{-1}R^{-1}Qy^\beta+\left(y^\beta\right)^TQD_\beta y$$

$$+\left(D_\beta y\right)^TRY\left(Y^\beta\right)^{-1}D_\beta y-\left(D_\beta y\right)^TRY\left(Y^\beta\right)^{-1}R^{-1}Qy$$

$$-y^TQY\left(Y^\beta\right)^{-1}D_\beta y+y^TQY\left(Y^\beta\right)^{-1}R^{-1}Qy$$

$$=\left(D_\beta y\right)^TQy-\left(y^\beta\right)^TPy^\beta-\left(y^\beta\right)^TQY\left(Y^\beta\right)^{-1}R^{-1}Qy^\beta$$

$$+\left(y^\beta\right)^TQD_\beta y+\left(D_\beta y\right)^TR\left(Y^\beta-(\beta-I)D_\beta Y\right)\left(Y^\beta\right)^{-1}D_\beta y$$

$$-\left(D_\beta y\right)^TR\left(Y^\beta-(\beta-I)D_\beta Y\right)\left(Y^\beta\right)^{-1}R^{-1}Qy$$

$$-y^TQY\left(Y^\beta\right)^{-1}D_\beta y+y^TQY\left(Y^\beta\right)^{-1}R^{-1}Qy$$

$$=\left(D_\beta y\right)^TRD_\beta y-\left(y^\beta\right)^TPy^\beta+\left(y^\beta\right)^TQD_\beta y$$

$$-(\beta-I)\left(D_\beta y\right)^TRD_\beta Y\left(Y^\beta\right)^{-1}D_\beta y$$

$$-y^TQY\left(Y^\beta\right)^{-1}D_\beta y+(\beta-I)\left(D_\beta y\right)^TRD_\beta Y\left(Y^\beta\right)^{-1}R^{-1}Qy$$

$$+y^TQY\left(Y^\beta\right)^{-1}R^{-1}Qy-\left(y^\beta\right)^TQY\left(Y^\beta\right)^{-1}R^{-1}Qy^\beta$$

$$
\begin{aligned}
&= (D_\beta y)^t R D_\beta y - \left(y^\beta\right)^T P y^\beta + \left(y^\beta\right)^T Q D_\beta y \\
&\quad - \left(y^\beta\right)^T QY \left(Y^\beta\right)^{-1} R^{-1} Q y^\beta - ((\beta - I) D_\beta y + y)^T QY \left(Y^\beta\right)^{-1} D_\beta y \\
&\quad + ((\beta - I) D_\beta y + y)^T QY \left(Y^\beta\right)^{-1} R^{-1} Q y \\
&= (D_\beta y)^T R D_\beta y - \left(y^\beta\right)^T P y^\beta + \left(y^\beta\right)^T Q D_\beta y \\
&\quad - \left(y^\beta\right)^T QY \left(Y^\beta\right)^{-1} R^{-1} Q y^\beta - \left(y^\beta\right)^T QY \left(Y^\beta\right)^{-1} D_\beta y \\
&\quad + \left(y^\beta\right)^T QY \left(Y^\beta\right)^{-1} R^{-1} Q y \\
&= (D_\beta y)^T R D_\beta y - \left(y^\beta\right)^T P y^\beta + \left(y^\beta\right)^T Q D_\beta y \\
&\quad - \left(y^\beta\right)^T Q \left(Y^\beta - (\beta - I) D_\beta Y\right) \left(Y^\beta\right)^{-1} D_\beta y \\
&\quad + \left(y^\beta\right)^T QY \left(Y^\beta\right)^{-1} R^{-1} Q \left(y - y^\beta\right) \\
&= (D_\beta y)^T R D_\beta y - \left(y^\beta\right)^T P y^\beta \\
&\quad + (\beta - I) \left(y^\beta\right)^T Q \left(D_\beta Y \left(Y^\beta\right)^{-1} - Y \left(Y^\beta\right)^{-1} D_\beta Y^{-1}\right) D_\beta y \\
&= (D_\beta y)^T R D_\beta y - \left(y^\beta\right)^T P y^\beta + \left(y^\beta\right)^T Q \left((\beta - I) D_\beta Y \left(Y^\beta\right)^{-1} \right. \\
&\quad \left. - Y \left(Y^\beta\right)^{-1} ((\beta - I) D_\beta Y) Y^{-1} \right) D_\beta y \\
&= (D_\beta y)^T R D_\beta y - \left(y^\beta\right)^T P y^\beta + \left(y^\beta\right)^T Q \left(\left(Y^\beta - Y\right) \left(Y^\beta\right)^{-1} \right. \\
&\quad \left. - Y \left(Y^\beta\right)^{-1} \left(Y^\beta - Y\right) Y^{-1} \right) D_\beta y \\
&= (D_\beta y)^T R D_\beta y - \left(y^\beta\right)^T P D_\beta y
\end{aligned}
$$

$$+\left(y^{\beta}\right)^{T} Q\left(I-Y\left(Y^{\beta}\right)^{-1}-I+Y\left(Y^{\beta}\right)^{-1}\right) D_{\beta} y$$

$$=\left(D_{\beta} y\right)^{T} R D_{\beta} y-\left(y^{\beta}\right)^{T} P y^{\beta}.$$

Therefore, at t we get

$$D_{\beta}\left(y^{T} Q y+2 \alpha^{T} Y^{-1} y-\alpha^{T} Y^{-1} Y_{1} \alpha\right)-\left(D_{\beta} y\right)^{T} R D_{\beta} y+\left(y^{\beta}\right)^{T} P y^{\beta}$$

$$=D_{\beta}\left(y^{T} Q y\right)-\left(D_{\beta} y\right)^{T} R D_{\beta} y+\left(y^{\beta}\right)^{T} P y^{\beta}$$

$$+2 \alpha^{T} D_{\beta}\left(Y^{-1} y\right)-\alpha^{T} D_{\beta}\left(Y^{-1} Y_{1}\right) \alpha$$

$$=-\left(R D_{\beta} y-Q y\right)^{T} Y\left(Y^{\beta}\right)^{-1} R^{-1}\left(R D_{\beta} y-Q y\right)$$

$$-2 \alpha^{T}\left(Y^{\beta}\right)^{-1} D_{\beta} Y Y^{-1} y+2 \alpha^{T}\left(Y^{\beta}\right)^{-1} D_{\beta} y$$

$$-\alpha^{T}\left(Y^{\beta}\right)^{-1} R^{-1}\left(Y^{-1}\right)^{T} \alpha$$

$$=-\left(R D_{\beta} y-Q y\right)^{T} Y\left(Y^{\beta}\right)^{-1} R^{-1}\left(R D_{\beta} y-Q y\right)$$

$$-2 \alpha^{T}\left(Y^{\beta}\right)^{-1} R^{-1} Q y+2 \alpha^{T}\left(Y^{\beta}\right)^{-1} D_{\beta} y$$

$$-\alpha^{T}\left(Y^{\beta}\right)^{-1} R^{-1}\left(Y^{-1}\right)^{T} \alpha$$

$$=-\left(R D_{\beta} y-Q y\right)^{T} Y\left(Y^{\beta}\right)^{-1} R^{-1}\left(R D_{\beta} y-Q y\right)$$

$$+2 \alpha^{T}\left(Y^{\beta}\right)^{-1} R^{-1}\left(R D_{\beta} y-Q y\right)$$

$$-\alpha^{T}\left(Y^{\beta}\right)^{-1} R^{-1}\left(Y^{-1}\right)^{T} \alpha$$

$$=-\left(R D_{\beta} y-Q y-\left(Y^{-1}\right)^{T} \alpha\right)^{T} Y\left(Y^{\beta}\right)^{-1} R^{-1}$$

$$\times\left(R D_{\beta} y-Q y-\left(Y^{-1}\right)^{T} \alpha\right).$$

This completes the proof. □

3.3 The Jacobi Condition

We will start this section with the following useful result.

Theorem 3.8 (Quadratic Functional)

Let $y \in \mathscr{C}^1_\beta([a,b])$, $y : [a,b] \to \mathbb{R}^n$, *and*

$$\left(D_\beta\left(RD_\beta y\right) + Py^\beta\right)(t) = 0, \quad t \in [a,b].$$

Then

$$\int_a^{\beta^{-1}(b)} \left(\left(D_\beta y\right)^T RD_\beta y - \left(y^\beta\right)^T Py^\beta\right)(t)D_\beta t = \left(y^T RD_\beta y\right)\left(\beta^{-1}(b)\right) - \left(y^T RD_\beta y\right)(a).$$

Proof On $[a,b]$ we have

$$\begin{aligned} D_\beta\left(y^T RD_\beta y\right) &= \left(y^\beta\right)^T D_\beta\left(RD_\beta y\right) + \left(D_\beta y\right)^T RD_\beta y \\ &= \left(y^\beta\right)^T\left(-Py^\beta\right) + \left(D_\beta y\right)^T RD_\beta y \\ &= \left(D_\beta y\right)^T RD_\beta y - \left(y^\beta\right)^T Py^\beta. \end{aligned}$$

Hence,

$$\begin{aligned} &\int_a^{\beta^{-1}(b)} \left(\left(D_\beta y\right)^T RD_\beta y - \left(y^\beta\right)^T Py^\beta\right)(t)D_\beta t \\ &= \int_a^{\beta^{-1}(b)} D_\beta\left(y^T RD_\beta y\right)(t)d_\beta t \\ &= \left(y^T RD_\beta y\right)(t)\Big|_{t=a}^{t=\beta^{-1}(b)} \\ &= \left(y^T RD_\beta y\right)\left(\beta^{-1}(b)\right) - \left(y^T RD_\beta y\right)(a). \end{aligned}$$

This completes the proof. □

We will consider a quadratic functional of the form

$$\mathscr{F}(y) = \int_a^b \left(\left(D_\beta y\right)^T RD_\beta y - \left(y^\beta\right)^T Py^\beta\right)(t)d_\beta t$$

for $y \in \mathscr{C}^1_\beta([a,b])$, $y : [a,b] \to \mathbb{R}^n$.

Definition 3.8 We will say that the functional $\mathscr{F}$ is positive definite and we will write $\mathscr{F} > 0$ if

$$\mathscr{F}(y) > 0 \quad for \quad all \quad nontrivial \quad y \in \mathscr{C}^1_\beta([a,b]), y:[a,b] \to \mathbb{R}^n.$$

Theorem 3.9 *A sufficient condition for $\mathscr{F} > 0$ is that there exist normalized conjoined solutions Y and Y_1 of (3.1) with Y invertible on $[a,b]$ and $Y\left(Y^\beta\right)^{-1}R^{-1}$ is positive definite on $[a,\beta^{-1}(b)]$.*

Proof Let $y \in \mathscr{C}^1_\beta([a,b])$, $y:[a,b] \to \mathbb{R}^n$, $y(a) = y(b) = 0$.

1. Suppose that a is right-scattered. Then $\beta(a) > a$ and $\beta(a) - a > 0$. Now, applying the Picone identity for $\alpha = 0$, we get

$$\begin{aligned}
\mathscr{F}(y) &= \int_a^b \left((D_\beta y)^T R D_\beta y - \left(y^\beta\right)^T P y^\beta \right)(t) d_\beta t \\
&= \int_a^{\beta(a)} \left((D_\beta y)^T R D_\beta y - \left(y^\beta\right)^T P y^\beta \right)(t) d_\beta t \\
&\quad + \int_{\beta(a)}^b \left((D_\beta y)^T R D_\beta y - \left(y^\beta\right)^T P y^\beta \right)(t) d_\beta t \\
&= \left((D_\beta y)^T R D_\beta y - \left(y^\beta\right)^T P y^\beta \right)(a)(\beta(a) - a) \\
&\quad + \int_{\beta(a)}^b \left((D_\beta y)^T R D_\beta y - \left(y^\beta\right)^T P y^\beta \right)(t) d_\beta t \\
&= \left((D_\beta y)^T R D_\beta y - \left(y^\beta\right)^T P y^\beta \right)(a)(\beta(a) - a) \\
&\quad + \int_{\beta(a)}^b \left(D_\beta\left(y^T Q y\right) + (R D_\beta y - Q y)^T D\left(R D_\beta y - Q y\right) \right)(t) d_\beta t \\
&= \left((D_\beta y)^T R D_\beta y - \left(y^\beta\right)^T P y^\beta \right)(a)(\beta(a) - a) \\
&\quad + \left(y^T Q y\right)(b) - \left(y^T Q y\right)(\beta(a)) \\
&\quad + \int_{\beta(a)}^b \left((R D_\beta y - Q y)^T D\left(R D_\beta y - Q y\right) \right)(t) d_\beta t \\
&\geq \left((D_\beta y)^T R D_\beta y - \left(y^\beta\right)^T P y^\beta \right)(a)(\beta(a) - a)
\end{aligned}$$

$$+ (y^T Qy)(b) - (y^T Qy)(\beta(a))$$

$$= \left(\left(y^\beta\right)^T Ry^\beta (\beta - I)^{-1} - \left(y^\beta\right)^T Py^\beta (\beta - I) - \left(y^\beta\right)^T (RD_\beta YY^{-1})^\beta y^\beta \right)(a)$$

$$= \left(y^\beta\right)^T \Big(R(\beta - I)^{-1} - P(\beta - I)$$

$$- \left(RY^\beta (\beta - I)^{-1} + (\beta - I) D_\beta (RD_\beta Y) \right) \left(Y^\beta\right)^{-1} \Big) y^\beta (a)$$

$$= \left(y^\beta\right)^T \left(R(\beta - I)^{-1} - P(\beta - I) - \left(RY^\beta (\beta - I)^{-1} - (\beta - I) PY^\beta \right) \left(Y^\beta\right)^{-1} \right) y^\beta (a)$$

$$= \left(y^\beta\right)^T (R(\beta - I)^{-1} - P(\beta - I) - R(\beta - I)^{-1} + P(\beta - I)) y^\beta (a)$$

$$= 0.$$

In particular, by the above computations, we get

$$\mathscr{F}(y) = \int_{\beta(a)}^{b} \left((RD_\beta y - Qy)^T D (RD_\beta y - Qy) \right)(t) d_\beta t.$$

Hence, if $\mathscr{F}(y) = 0$, using that P is a positive definite matrix, we conclude that

$$(RD_\beta y - Qy)(t) = 0, \quad t \in [\beta(a), \beta^{-1}(b)],$$

or

$$(RD_\beta y)(t) = (Qy)(t), \quad t \in [\beta(a), \beta^{-1}(b)],$$

or

$$D_\beta y(t) = \left(D_\beta YY^{-1} y\right)(t), \quad t \in [\beta(a), \beta^{-1}(b)].$$

Note that

$$\begin{aligned} 1 + (\beta - I) D_\beta YY^{-1} &= I + \left(Y^\beta - Y\right) Y^{-1} \\ &= I + Y^\beta Y^{-1} - I \\ &= Y^\beta Y^{-1} \\ &\neq 0. \end{aligned}$$

Therefore, $D_\beta YY^{-1}$ is a regressive matrix. Then the IVP

$$D_\beta y(t) = \left(D_\beta YY^{-1} y\right)(t), \quad t \in [\beta(a), \beta^{-1}(b)],$$

$$y(b) = 0,$$

has unique solution $y(t) = 0$, $t \in [a, b]$. Therefore, $\mathscr{F} > 0$.

2. Let a be right-dense. Then there exists a right-dense sequence $\{t_m\}_{m\in\mathbb{N}} \subset [a,b]$ such that $t_m \to a$, $m \to \in fty$. Let

$$\alpha_m = -\left(\left(D_\beta Y\right)^T Ry\right)(t_m), \quad m \in \mathbb{N}.$$

Now we apply the Picone identity for $\alpha = \alpha_m$ and we get

$$\int_{t_m}^{b} \left(\left(D_\beta y\right)^T RD_\beta y - \left(y^\beta\right)^T Py^\beta\right)(t)d_\beta t$$

$$= \int_{t_m}^{b} D_\beta \left(y^T Qy + 2\alpha_m^T Y^{-1} y - \alpha_m^T Y^{-1} Y_1 \alpha_m\right)(t) d_\beta t$$

$$+ \int_{t_m}^{b} \left(\left(RD_\beta y - Qy - \left(Y^{-1}\right)^T \alpha_m\right)^T D\left(RD_\beta y - Qy - \left(Y^{-1}\right)^T \alpha_m\right)\right)(t)d_\beta t$$

$$\geq \int_{t_m}^{b} D_\beta \left(y^T Qy + 2\alpha_m^T Y^{-1} y - \alpha_m^T Y^{-1} Y_1 \alpha_m\right)(t) d_\beta t$$

$$= \left(y^T Qy + 2\alpha_m^T Y^{-1} y - \alpha_m^T Y^{-1} Y_1 \alpha_m\right)\Big|_{t=t_m}^{t=b}$$

$$= -\left(\alpha_m^T Y^{-1} Y_1 \alpha_m\right)(b)$$

$$- \left(y^T Qy + 2\alpha_m^T Y^{-1} y - \alpha_m^T Y^{-1} Y_1 \alpha_m\right)(t_m)$$

$$= -\left(\alpha_m^T Y^{-1} Y_1 \alpha_m\right)(b)$$

$$+ \left(-y^T Qy + 2y^T RD_\beta Y Y^{-1} y + y^T RD_\beta Y Y^{-1} Y_1 \left(D_\beta Y\right)^T Ry\right)(t_m)$$

$$= -\left(\alpha_m^T Y^{-1} Y_1 \alpha_m\right)(b)$$

$$+ \left(y^T Qy + y^T QY_1 \left(D_\beta Y\right)^T Ry\right)(t_m). \tag{3.7}$$

Since Y and Y_1 are normalized, we have

$$Y^T RD_\beta Y_1 - I = \left(D_\beta Y\right)^T RY_1$$

$$= \left(D_\beta Y\right)^T RYY^{-1} Y_1.$$

Because Y is conjoined, we have

$$(D_\beta Y)^T RY = Y^T RD_\beta Y.$$

Therefore,

$$Y^T RD_\beta Y_1 - I = Y^T RD_\beta YY^{-1}Y_1$$

$$= Y^T QY_1$$

and then

$$QY_1 = \left(Y^T\right)^{-1} \left(Y^T RD_\beta Y_1 - I\right).$$

From this and (3.7), using that Q is symmetric, we obtain

$$\int_{t_m}^{b} \left((D_\beta y)^T RD_\beta y - \left(y^\beta\right)^T Py^\beta \right)(t) d_\beta t = -\left(\alpha_m^T Y^{-1} Y_1 \alpha_m\right)(b)$$

$$+ \left(y^T Qy + y^T \left(Y^T\right)^{-1} \left(Y^T RD_\beta Y_1 - I\right) (D_\beta Y)^T Ry \right)(t_m)$$

$$= -\left(\alpha_m^T Y^{-1} Y_1 \alpha_m\right)(b)$$

$$+ \left(y^T Qy + y^T \left(RD_\beta Y_1 (D_\beta Y)^T R - \left(Y^T\right)^{-1} (D_\beta Y)^T R \right) y \right)(t_m)$$

$$= -\left(\alpha_m^T Y^{-1} Y_1 \alpha_m\right)(b)$$

$$+ \left(y^T Qy + y^T RD_\beta Y_1 (D_\beta Y)^T Ry - y^T Qy \right)(t_m)$$

$$= -\left(\alpha_m^T Y^{-1} Y_1 \alpha_m\right)(b) + \left(y^T RD_\beta Y_1 (D_\beta Y)^T Ry \right)(t_m).$$

Consequently,

$$\mathscr{F}(y) = \lim_{m \to \in fty} \int_{t_m}^{b} \left((D_\beta y)^T RD_\beta y - \left(y^\beta\right)^T Py^\beta \right)(t) d_\beta t$$

$$\geq - \lim_{m \to \in fty} \left(\alpha_m^T \left(Y^{-1} Y_1\right)(b) \alpha_m + y^T \left(RD_\beta Y_1\right)(t_m) \alpha_m \right)$$

$$= - \lim_{m \to \in fty} \left(\left(y^T RD_\beta Y\right)(t_m) \left(Y^{-1} Y_1\right)(b) \alpha_m + y^T \left(RD_\beta Y_1\right)(t_m) \alpha_m \right)$$

$$= 0.$$

Now we suppose that $\mathscr{F}(y) = 0$. Let

$$z_m = RD_\beta y - Qy - \left(Y^{-1}\right)^T \alpha_m \to RD_\beta y - Qy = z, \quad as \quad m \to \in fty,$$

uniformly. Let $k \in \mathbb{N}$ and $m \geq k$. Then

$$\int_{t_m}^{b} \left(z_m^T D z_m\right)(t) d_\beta t \geq \int_{t_k}^{b} \left(z_m^T D z_m\right)(t) d_\beta t.$$

From here,

$$\begin{aligned} 0 &= \lim_{m \to \in fty} \int_{t_m}^{b} \left(z_m^T D z_m\right)(t) d_\beta t \\ &\geq \int_{t_k}^{b} \left(z^T D z\right)(t) d_\beta t \end{aligned}$$

and

$$\begin{aligned} 0 &= \lim_{k \to \in fty} \left(\lim_{m \to \in fty} \int_{t_m}^{b} \left(z_m^T D z_m\right)(t) d_\beta t \right) \\ &\geq \lim_{k \to \in fty} \int_{t_k}^{b} \left(z^T D z\right)(t) d_\beta t \\ &\geq 0. \end{aligned}$$

Therefore, $z = 0$ and hence, $y = 0$. We conclude that $\mathscr{F}$ is positive definite. This completes the proof. □

Definition 3.9 We say that equation (3.1) is disconjugate on $[a, b]$ if the principal solution $\tilde{Y}$ of (3.1) satisfies

$$\tilde{Y} \quad is \quad invertible \quad on \quad [a,b], \quad \tilde{Y}\left(\tilde{Y}^\beta\right)^{-1} R^{-1} \quad is \quad positive \quad definite \quad on \quad [a,b].$$

Theorem 3.10 (The Jacobi Condition)

$\mathscr{F} > 0$ *if and only if* (3.1) *is disconjugate.*

Proof 1. Let (3.1) be disconjugate. Let $\tilde{Y}$ is the principal solution of (3.1) and $\tilde{Y}_1$ is the associated solution of (3.1). Then $\tilde{Y}$ and $\tilde{Y}_1$ are normalized. We apply Theorem 3.9 and we conclude that $\mathscr{F} > 0$.

2. Let $\mathscr{F} > 0$. Assume that (3.1) is not disconjugate. Then there exists a point $\tilde{t_0} \in [a,b]$ with exactly one of the following properties:

Case 1. $\tilde{t_0} \in (a,b]$, $\tilde{Y}(t)$ is invertible on $(a,\tilde{t_0})$ and $\tilde{Y}(\tilde{t_0})$ is singular.
Case 2. $\tilde{t_0} \in (a,b]$ such that $\tilde{Y}(t)$ is invertible for all $t \in (a,b)$ and

$$\tilde{D}(\tilde{t_0}) = \left(\tilde{Y}\left(\tilde{Y}^{\beta}\right)^{-1} R^{-1}\right)(\tilde{t_0})$$

is not positive definite.

Let $d \in \mathbb{R}^n \backslash \{0\}$ be such that

$$\tilde{Y}(\tilde{t_0})d = 0$$

in case 1, and

$$\left(d^T \tilde{D} d\right)(\tilde{t_0}) \leq 0$$

in case 2. We set

$$y(t) = \begin{cases} \tilde{Y}(t)d & for \quad t \leq \tilde{t_0} \\ 0 & otherwise. \end{cases}$$

Since $\tilde{Y}$ is the principal solution, we have

$$\begin{aligned} y(a) &= \tilde{Y}(a)d \\ &= 0, \end{aligned}$$

and using the definition of y, we have

$$y(b) = 0.$$

Also,

$$y(t) \neq 0 \quad for \quad all \quad t \in (a,\tilde{t_0}).$$

Note that $y \in \mathscr{C}_p^1([a,b])$ except for case 2 with $\tilde{t_0}$ right-dense. We have that $y^{D_\beta}(t)$ and $y^{\beta}(t)$ are zero for all $t > \tilde{t_0}$. Hence,

$$\left(\left(D_\beta y\right)^T R D_\beta y - \left(y^\beta\right)^T P y^\beta\right)(t) = 0 \quad for \quad all \quad t > \tilde{t_0}.$$

Using Theorem 3.8, we obtain

$$
\begin{aligned}
\mathscr{F}(y) &= \int_a^{\beta(\widetilde{t_0})} \left((D_\beta y)^T R D_\beta y - \left(y^\beta\right)^T P y^\beta \right)(t) d_\beta t \\
&= \int_a^{\widetilde{t_0}} \left((D_\beta y)^T R D_\beta y - \left(y^\beta\right)^T P y^\beta \right)(t) d_\beta t \\
&\quad + \int_{\widetilde{t_0}}^{\beta(\widetilde{t_0})} \left((D_\beta y)^T R D_\beta y - \left(y^\beta\right)^T P y^\beta \right)(t) d_\beta t \\
&= \left(y^T R D_\beta y\right)(\widetilde{t_0}) + \left((D_\beta y)^T R D_\beta y - \left(y^\beta\right)^T P y^\beta \right)(\widetilde{t_0})\mu(\widetilde{t_0}) \\
&= \left(y^T R D_\beta y\right)(\widetilde{t_0}) + D_\beta \left(y^T R D_\beta y\right)(\widetilde{t_0})(\beta(\widetilde{t_0}) - \widetilde{t_0}) \\
&= \left(d^T \tilde{Y}^T R D_\beta \tilde{Y} d\right)(\widetilde{t_0}) + \left(y^T R D_\beta y\right)^\beta (\widetilde{t_0}) - \left(y^T R D_\beta y\right)(\widetilde{t_0}) \\
&= \left(d^T \tilde{Y}^T R D_\beta \tilde{Y} d\right)(\widetilde{t_0}) - \left(y^T R D_\beta y\right)(\widetilde{t_0}) \\
&= 0.
\end{aligned}
$$

This is a contradiction. Now we consider case 2 when $\widetilde{t_0} = t_0$. There exists $d \in \mathbb{R}^n$ such that

$$d^T R(t_0) d < 0.$$

Assume that $\mathscr{F} > 0$. For $m \in \mathbb{N}$, we consider

$$
y_m(t) = \begin{cases} \dfrac{t - t_m^*}{\sqrt{\widetilde{t_0} - t_m^*}} d & if \quad t \in [t_m^*, \widetilde{t_0}) \\[2ex] \dfrac{t_m - t}{\sqrt{t_m - \widetilde{t_0}}} d & if \quad t \in (\widetilde{t_0}, t_m] \\[2ex] 0 \quad otherwise, \end{cases}
$$

if $\widetilde{t_0} \neq b$, and

$$
y_m(t) = \begin{cases} \dfrac{t - t_m^*}{\sqrt{\widetilde{t_0} - t_m^*}} d & if \quad t \in [t_m^*, \widetilde{t_0}) \\[2ex] 0 \quad otherwise, \end{cases}
$$

if $\widetilde{t_0} = b$, where $\{t_m^*\}_{m\in\mathbb{N}} \subset [a,b]$ is a strictly increasing sequence with $\lim_{m\to\in fty} t_m^* = \widetilde{t_0}$. Again we assume $\mathscr{F} > 0$. Then, if $\widetilde{t_0} \neq b$, we have

$$
\begin{aligned}
0 &< \mathscr{F}(y_m) \\
&= \int_{\beta(t_m^*)}^{\widetilde{t_0}} \left(\left(y^{D_\beta}\right)^T R y^{D_\beta} - \left(y^\beta\right)^T P y^\beta \right)(t) D_\beta t \\
&\quad + \int_{\widetilde{t_0}}^{\beta(t_m)} \left(\left(y^{D_\beta}\right)^T R y^{D_\beta} - \left(y^\beta\right)^T P y^\beta \right)(t) D_\beta t \\
&= \int_{\beta(t_m^*)}^{\widetilde{t_0}} \left(\frac{d^T}{\sqrt{\widetilde{t_0} - t_m^*}} R(t) \frac{d}{\sqrt{\widetilde{t_0} - t_m^*}} - d^T \frac{\beta(t) - t_m^*}{\sqrt{\widetilde{t_0} - t_m^*}} P(t) \frac{\beta(t) - t_m^*}{\sqrt{\widetilde{t_0} - t_m^*}} d \right) D_\beta t \\
&\quad + \int_{\widetilde{t_0}}^{\beta(t_m)} \left(\frac{d^T}{\sqrt{t_m - \widetilde{t_0}}} R(t) \frac{d}{\sqrt{t_m - \widetilde{t_0}}} - d^T \frac{t_m - \beta(t)}{\sqrt{t_m - \widetilde{t_0}}} P(t) \frac{t_m - \beta(t)}{\sqrt{t_m - \widetilde{t_0}}} d \right) D_\beta t \\
&\to 2 d^T R(\widetilde{t_0}) d < 0, \quad m \to \in fty,
\end{aligned}
$$

and, if $\widetilde{t_0} \neq b$, then

$$
\begin{aligned}
0 &< \mathscr{F}(y_m) \\
&= \int_{\beta(t_m^*)}^{\widetilde{t_0}} \left(\left(y^{D_\beta}\right)^T R y^{D_\beta} - \left(y^\beta\right)^T P y^\beta \right)(t) D_\beta t \\
&= \int_{\beta(t_m^*)}^{\widetilde{t_0}} \left(\frac{d^T}{\sqrt{\widetilde{t_0} - t_m^*}} R(t) \frac{d}{\sqrt{\widetilde{t_0} - t_m^*}} - d^T \frac{\beta(t) - t_m^*}{\sqrt{\widetilde{t_0} - t_m^*}} P(t) \frac{\beta(t) - t_m^*}{\sqrt{\widetilde{t_0} - t_m^*}} d \right) D_\beta t \\
&\to d^T R(\widetilde{t_0}) d < 0, \quad m \to \in fty,
\end{aligned}
$$

i.e., we go to a contradiction. This completes the proof. □

3.4 Sturmian Theory

Definition 3.10 We call a solution Y of equation (3.1) basis whenever

$$rank\begin{pmatrix} Y(a) \\ D_\beta Y(a) \end{pmatrix} = n.$$

Definition 3.11 A conjoined solution Y of (3.1) is said to have no focal points in $(a,b]$ if Y is invertible on $(a,b]$ and $Y\left(Y^\beta\right)^{-1}R^{-1}$ is positive definite on $(a,b]$.

By Definition 3.11, it follows that equation (3.1) is disconjugate on $[a,b]$ if and only if the principal solution of (3.1) at a has no focal points in $(a,b]$.

Theorem 3.11 (The Sturm Separation Theorem)

Suppose there exists a conjoined basis of (3.1) *with no focal points in* $(a,b]$. *Then* (3.1) *is disconjugate in* $[a,b]$.

Proof Let Y be a conjoined basis of (3.1) such that Y is invertible in $(a,b]$ and $Y\left(Y^\beta\right)^{-1}R^{-1}$ is positive definite on $(a,\beta^{-1}(b)]$. Note that

$$K = Y^T(a)Y(a) + \left(D_\beta Y\right)(a)R^2(a)D_\beta Y(a)$$

is invertible because Y is a basis. Let Y_1 be the solution of (3.1) such that

$$Y_1(a) = -R(a)D_\beta Y(a)K^{-1}, \quad D_\beta Y_1(a) = R^{-1}(a)Y(a)K^{-1}.$$

By the Wronskian identity, we have

$$\left(Y_1^T RD_\beta Y_1 - \left(D_\beta Y_1\right)^T RY_1\right)(t) = \left(Y_1^T RD_\beta Y_1 - \left(D_\beta Y_1\right)^T RY_1\right)(a)$$

$$= -\left(K^{-1}\right)^T\left(D_\beta Y\right)^T(a)R(a)R(a)R^{-1}(a)Y(a)K^{-1}$$

$$+\left(K^{-1}\right)^T Y^T(a)\left(R^{-1}\right)^T(a)R(a)R(a)D_\beta Y(a)K^{-1}$$

$$= -\left(K^{-1}\right)^T\left(D_\beta Y\right)^T(a)R(a)Y(a)K^{-1}$$

$$+\left(K^{-1}\right)^T Y^T(a)R(a)D_\beta Y(a)K^{-1}$$

$$= \left(K^{-1}\right)^T\left(Y^T(a)R(a)D_\beta Y(a) - \left(D_\beta Y\right)^T(a)R(a)Y(a)\right)K^{-1}$$

$$= 0$$

and

$$\left(Y^T RD_\beta Y_1 - (D_\beta Y)^T RY_1\right)(t) = \left(Y^T RD_\beta Y_1 - (D_\beta Y)^T RY_1\right)(a)$$

$$= Y^T(a)R(a)R^{-1}(a)Y(a)K^{-1} + (D_\beta Y)^T(a)R^2(a)D_\beta Y(a)K^{-1}$$

$$= Y^T(a)Y(a)K^{-1} + (D_\beta Y)^T(a)R^2(a)D_\beta Y(a)K^{-1}$$

$$= KK^{-1}$$

$$= I.$$

Therefore, Y and Y_1 are normalized solutions of equation (3.1). Hence, applying Theorem 3.9, we get $\mathscr{F} > 0$. From here and from Theorem 3.10, we obtain that (3.1) is disconjugate, i.e., the principal solution of equation (3.1) has no focal points in $(a,b]$. This completes the proof. □

Now we consider the equation

$$D_\beta\left(\tilde{R}(t)D_\beta Y\right) + \tilde{P}(t)Y^\beta = 0, \tag{3.8}$$

where $\tilde{R}$ and $\tilde{P}$ satisfy the same assumptions as R and P.

Theorem 3.12 (The Sturm Comparison Theorem)

Let

$$\tilde{R}(t) \le R(t) \quad and \quad \tilde{P}(t) \ge P(t), \quad t \in [a,b].$$

If (3.8) *is disconjugate, then* (3.1) *is disconjugate.*

Proof Let (3.8) be disconjugate. Then, by the Jacobi condition, Theorem 3.10, we obtain

$$\tilde{\mathscr{F}}(y) = \int_a^b \left((D_\beta y)^T \tilde{R} D_\beta y - \left(y^\beta\right)^T \tilde{P} y^\beta\right)(t)d_\beta t$$

$$> 0$$

for all nontrivial $y \in \mathscr{C}_p^1([a,b])$ with $y(a) = y(b) = 0$. For such y we also have

$$\mathscr{F}(y) = \int_a^b \left((D_\beta y)^T R D_\beta y - \left(y^\beta\right)^T P y^\beta\right)(t)d_\beta t$$

$$\ge \int_a^b \left((D_\beta y)^T \tilde{R} D_\beta y - \left(y^\beta\right)^T \tilde{P} y^\beta\right)(t)d_\beta t$$

$$> 0,$$

i.e., $\mathscr{F} > 0$ and equation (3.1) is disconjugate. This completes the proof. □

Chapter 4
Linear Hamiltonian Dynamic Systems

Let $I \subset \mathbb{R}$ and β be a general first kind or second kind quantum operator.

Throughout this chapter, we denote by $\mathscr{I}$ the $2n \times 2n$ matrix

$$\mathscr{I} = \begin{pmatrix} O & I \\ -I & O \end{pmatrix}.$$

4.1 Linear Symplectic Dynamic Systems

Definition 4.1 (Symplectic Matrix)

A $2n \times 2n$ matrix A is called symplectic if

$$A^* \mathscr{I} A = \mathscr{I}.$$

Example 4.1 Consider the matrix

$$A = \begin{pmatrix} 2 & 0 & 2 & 0 \\ 0 & 1 & 0 & 2 \\ 3 & 0 & \frac{7}{2} & 0 \\ 0 & 3 & 0 & 7 \end{pmatrix}.$$

Then

$$\mathscr{J}A = \begin{pmatrix} 0 & 0 & 1 & 0 \\ 0 & 0 & 0 & 1 \\ -1 & 0 & 0 & 0 \\ 0 & -1 & 0 & 0 \end{pmatrix} \begin{pmatrix} 2 & 0 & 2 & 0 \\ 0 & 1 & 0 & 2 \\ 3 & 0 & \frac{7}{2} & 0 \\ 0 & 3 & 0 & 7 \end{pmatrix}$$

$$= \begin{pmatrix} 3 & 0 & \frac{7}{2} & 0 \\ 0 & 3 & 0 & 7 \\ -2 & 0 & -2 & 0 \\ 0 & -1 & 0 & -2 \end{pmatrix},$$

$$A^* = \begin{pmatrix} 2 & 0 & 3 & 0 \\ 0 & 1 & 0 & 3 \\ 2 & 0 & \frac{7}{2} & 0 \\ 0 & 2 & 0 & 7 \end{pmatrix},$$

$$A^*JA = \begin{pmatrix} 2 & 0 & 3 & 0 \\ 0 & 1 & 0 & 3 \\ 2 & 0 & \frac{7}{2} & 0 \\ 0 & 2 & 0 & 7 \end{pmatrix} \begin{pmatrix} 3 & 0 & \frac{7}{2} & 0 \\ 0 & 3 & 0 & 7 \\ -2 & 0 & -2 & 0 \\ 0 & -1 & 0 & -2 \end{pmatrix}$$

$$= \begin{pmatrix} 0 & 0 & 1 & 0 \\ 0 & 0 & 0 & 1 \\ -1 & 0 & 0 & 0 \\ 0 & -1 & 0 & 0 \end{pmatrix}.$$

Therefore, A is symplectic.

Theorem 4.1 *Let A, B, C and D be $n \times n$ matrices. Then the matrix*

$$\begin{pmatrix} A & B \\ C & D \end{pmatrix}$$

is symplectic if and only if A^*C and B^*D are Hermitian and

$$A^*D - C^*B = I.$$

Proof The matrix

$$\begin{pmatrix} A & B \\ C & D \end{pmatrix}$$

is symplectic if and only if

$$\mathscr{I} = \begin{pmatrix} A^* & C^* \\ B^* & D^* \end{pmatrix} \begin{pmatrix} O & I \\ -I & O \end{pmatrix} \begin{pmatrix} A & B \\ C & D \end{pmatrix} \quad iff$$

$$\begin{pmatrix} O & I \\ -I & O \end{pmatrix} = \begin{pmatrix} A^* & C^* \\ B^* & D^* \end{pmatrix} \begin{pmatrix} C & D \\ -A & -B \end{pmatrix}$$

$$= \begin{pmatrix} A^*C - C^*A & A^*D - C^*B \\ B^*C - D^*A & B^*D - D^*B \end{pmatrix} \quad iff$$

$$A^*C = C^*A, \quad B^*D = D^*B, \quad A^*D - C^*B = I.$$

This completes the proof. □

Theorem 4.2 *Let A, B, C and D be $n \times n$ matrices. Then*

$$\begin{pmatrix} A & B \\ C & D \end{pmatrix}$$

is symplectic if and only if it is invertible with

$$\begin{pmatrix} A & B \\ C & D \end{pmatrix}^{-1} = \begin{pmatrix} D^* & -B^* \\ -C^* & A^* \end{pmatrix}.$$

Proof We have

$$\begin{pmatrix} D^* & -B^* \\ -C^* & A^* \end{pmatrix} \begin{pmatrix} A & B \\ C & D \end{pmatrix} = \begin{pmatrix} D^*A - B^*C & D^*B - B^*D \\ -C^*A + A^*C & -C^*B + A^*D \end{pmatrix}.$$

Hence, using Theorem 4.1, we conclude that

$$\begin{pmatrix} A & B \\ C & D \end{pmatrix}$$

is symplectic if and only if

$$A^*C = C^*A, \quad B^*D = D^*B, \quad A^*D - C^*B = I, \quad D^*A - B^*C = I,$$

if and only if

$$\begin{pmatrix} D^*A - B^*C & D^*B - B^*D \\ -C^*A + A^*C & -C^*B + A^*D \end{pmatrix} = \begin{pmatrix} I & O \\ O & I \end{pmatrix}$$

if and only if

$$\begin{pmatrix} D^* & -B^* \\ -C^* & A^* \end{pmatrix} \begin{pmatrix} A & B \\ C & D \end{pmatrix} = \begin{pmatrix} I & O \\ O & I \end{pmatrix}.$$

This completes the proof. □

Exercise 4.1 Let A, B, C and D be $n \times n$ matrices. Prove that the matrix

$$\begin{pmatrix} A & B \\ C & D \end{pmatrix}$$

is symplectic if and only if the matrices AB^* and CD^* are Hermitian and

$$AD^* - BC^* = I.$$

Hint. Use that

$$\begin{pmatrix} I & O \\ O & I \end{pmatrix} = \begin{pmatrix} A & B \\ C & D \end{pmatrix} \begin{pmatrix} D^* & -B^* \\ -C^* & A^* \end{pmatrix}$$
$$= \begin{pmatrix} AD^* - BC^* & BA^* - AB^* \\ CD^* - DC^* & -CB^* + DA^* \end{pmatrix}$$

and then apply Theorem 4.2.

Definition 4.2 A $2n \times 2n$ matrix-valued function S is called symplectic with respect to I if

$$S^*(t)\mathscr{J} + \mathscr{J}S(t) + (\beta(t) - t)S^*(t)JS(t) = O, \quad t \in I.$$

If S is symplectic, then the system

$$z^{D_\beta} = S(t)z \tag{4.1}$$

is called a linear symplectic dynamic system.

Theorem 4.3 *If S is symplectic with respect to I, then $I + \mu S$ is symplectic.*

Proof We have

$$\begin{aligned}
&(I + (\beta - I)S)^* \mathscr{J} (I + (\beta - I)S) \\
&\quad = (I + (\beta - I)S^*) \mathscr{J} (I + (\beta - I)S) \\
&\quad = (I + (\beta - I)S^*)(\mathscr{J} + (\beta - I)JS) \\
&\quad = \mathscr{J} + (\beta - I)\mathscr{J}S + (\beta - I)S^* \mathscr{J} + (\beta - I)^2 S^* \mathscr{J} S \\
&\quad = \mathscr{J} + (\beta - I)(\mathscr{J}S + S^* \mathscr{J} + (\beta - I)S^* \mathscr{J} S) \\
&\quad = \mathscr{J}.
\end{aligned}$$

This completes the proof. $\square$

Theorem 4.4 *If S is symplectic with respect to I, then it is regressive.*

Proof Since S is symplectic with respect to I, then $I+(\beta-I)S$ is symplectic. From this and, it follows that $I+(\beta-I)S$ is invertible. Therefore, S is regressive. This completes the proof. □

Corollary 4.1 *Let $S \in \mathscr{C}_\beta$ is symplectic with respect to I, $t_0 \in I$, $z_0 \in \mathbb{R}^{2n}$. Then the initial value problem (IVP)*

$$D_\beta z = S(t)z, \quad z(t_0) = z_0,$$

has unique solution $z : I \to \mathbb{R}^{2n}$.

Theorem 4.5 *Let A, B, C and D be $n \times n$ matrices and*

$$S = \begin{pmatrix} A & B \\ C & D \end{pmatrix}.$$

Then S is symplectic with respect to I if and only if

$$(I+(\beta-I)A)^*C - C^*(I+(\beta-I)A) = O,$$

$$(I+(\beta-I)A)^* + A^* - (\beta-I)C^*B = O,$$

$$B^*(I+(\beta-I)D) - (I+(\beta-I)D)^*B = O.$$

Proof We have

$$S^* = \begin{pmatrix} A^* & C^* \\ B^* & D^* \end{pmatrix}$$

and

$$\begin{aligned}
S^*J + JS + (\beta-I)S^*JS &= \begin{pmatrix} A^* & C^* \\ B^* & D^* \end{pmatrix}\begin{pmatrix} O & I \\ -I & O \end{pmatrix} + \begin{pmatrix} O & I \\ -I & O \end{pmatrix}\begin{pmatrix} A & B \\ C & D \end{pmatrix} \\
&\quad + (\beta-I)\begin{pmatrix} A^* & C^* \\ B^* & D^* \end{pmatrix}\begin{pmatrix} O & I \\ -I & O \end{pmatrix}\begin{pmatrix} A & B \\ C & D \end{pmatrix} \\
&= \begin{pmatrix} -C^* & A^* \\ -D^* & B^* \end{pmatrix} + \begin{pmatrix} C & D \\ -A & -B \end{pmatrix} \\
&\quad + (\beta-I)\begin{pmatrix} A^* & C^* \\ B^* & D^* \end{pmatrix}\begin{pmatrix} C & D \\ -A & -B \end{pmatrix} \\
&= \begin{pmatrix} C-C^* & A^*+D \\ -A-D^* & B^*-B \end{pmatrix} \\
&\quad + (\beta-I)\begin{pmatrix} A^*C - C^*A & A^*D - C^*B \\ B^*C - D^*A & B^*D - D^*B \end{pmatrix}
\end{aligned}$$

$$= \begin{pmatrix} C-C^* & D+A^* \\ -(D+A^*)^* & B^*-B \end{pmatrix}$$

$$+(\beta - I)\begin{pmatrix} A^*C-C^*A & A^*D-C^*B \\ -(A^*D-C^*B)^* & B^*D-D^*B \end{pmatrix}$$

$$= \begin{pmatrix} C+(\beta-I)A^*C-(\beta-I)C^*A-C^* & D+(\beta-I)A^*D+A^*-(\beta-I)C^*B \\ -(A^*+D+(\beta-I)A^*D-(\beta-I)C^*B)^* & B^*-B+(\beta-I)B^*D-(\beta-I)D^*B \end{pmatrix}$$

$$= \begin{pmatrix} (I+(\beta-I)A)^*C-C^*(I+(\beta-I)A) & (I+(\beta-I)A)^*D+A^*-(\beta-I)C^*B \\ -((I+(\beta-I)A^*)D+A^*-(\beta-I)C^*B)^* & B^*(I+(\beta-I)D)-(I+(\beta-I)D)^*B \end{pmatrix}.$$

From here, we conclude that S is symplectic with respect to I if and only if

$$(I+(\beta-I)A)^*C-C^*(I+(\beta-I)A)=0,$$

$$(I+(\beta-I)A)^*+A^*-(\beta-I)C^*B=0,$$

$$B^*(I+(\beta-I)D)-(I+(\beta-I)D)^*B=0.$$

This completes the proof. □

4.2 Hamiltonian Systems

Definition 4.3 A $2n\times 2n$ matrix A is called Hamiltonian if

$$A^*\mathscr{J}+\mathscr{J}A=O.$$

Theorem 4.6 *Let A, B, C and D be $n\times n$ matrices. Then the matrix*

$$\begin{pmatrix} A & B \\ C & D \end{pmatrix}$$

is Hamiltonian if and only if B and C are Hermitian and $D=-A^*$.

Proof We have

$$\begin{pmatrix} A & B \\ C & D \end{pmatrix}^* = \begin{pmatrix} A^* & C^* \\ B^* & D^* \end{pmatrix}$$

and

$$\begin{pmatrix} A & B \\ C & D \end{pmatrix}^*\mathscr{J}+\mathscr{J}\begin{pmatrix} A & B \\ C & D \end{pmatrix} = \begin{pmatrix} A^* & C^* \\ B^* & D^* \end{pmatrix}\begin{pmatrix} O & I \\ -I & O \end{pmatrix}+\begin{pmatrix} O & I \\ -I & O \end{pmatrix}\begin{pmatrix} A & B \\ C & D \end{pmatrix}$$

$$= \begin{pmatrix} -C^* & A^* \\ -D^* & B^* \end{pmatrix}+\begin{pmatrix} C & D \\ -A & -B \end{pmatrix}$$

$$= \begin{pmatrix} -C^*+C & A^*+D \\ -D^*-A & B^*-B \end{pmatrix}.$$

Hence,

$$\begin{pmatrix} O & O \\ O & O \end{pmatrix} = \begin{pmatrix} -C^* + C & A^* + D \\ -D^* - A & B^* - B \end{pmatrix}$$

if and only if

$$C^* = C, \quad B^* = B \quad and \quad A^* = -D.$$

This completes the proof. □

Definition 4.4 A $2n \times 2n$ matrix-valued function $\mathscr{H}$ is called Hamiltonian with respect to I if $\mathscr{H}$ is Hamiltonian and the matrix

$$I - (\beta(t) - t)\mathscr{H}(t)\mathscr{M}^*\mathscr{M}$$

is invertible for all $t \in I$, where

$$\mathscr{M} = \begin{pmatrix} O & O \\ I & O \end{pmatrix}.$$

In this case, the system

$$D_\beta z = \mathscr{H}(t)\left(\mathscr{M}^*\mathscr{M}z^\beta + \mathscr{M}\mathscr{M}^*z\right) \tag{4.2}$$

is called a linear Hamiltonian dynamic system.

Remark 4.1 Let

$$\mathscr{H} = \begin{pmatrix} A & B \\ C & D \end{pmatrix},$$

where A, B, C and D are $n \times n$ matrices. Then

$$\begin{aligned} I - (\beta(t) - t)\mathscr{H}(t)\mathscr{M}^*\mathscr{M} &= \begin{pmatrix} I & O \\ O & I \end{pmatrix} - (\beta(t) - t)\begin{pmatrix} A & B \\ C & D \end{pmatrix} \\ &\quad \times \begin{pmatrix} O & I \\ O & O \end{pmatrix}\begin{pmatrix} O & O \\ I & O \end{pmatrix} \\ &= \begin{pmatrix} I & O \\ O & I \end{pmatrix} - (\beta(t) - t)\begin{pmatrix} A & B \\ C & D \end{pmatrix}\begin{pmatrix} I & O \\ O & O \end{pmatrix} \\ &= \begin{pmatrix} I & O \\ O & I \end{pmatrix} - (\beta(t) - t)\begin{pmatrix} A & O \\ C & O \end{pmatrix} \\ &= \begin{pmatrix} I - (\beta(t) - t)A & O \\ -(\beta(t) - t)C & I \end{pmatrix}. \end{aligned}$$

From this and, $\mathscr{H}$ is Hamiltonian with respect to I if and only if

$$B^* = B, \quad C^* = C, \quad D = -A^*$$

and $I - \mu A$ is invertible for all $t \in I$. Now we suppose that

$$\mathscr{H} = \begin{pmatrix} A & B \\ -C & D \end{pmatrix}$$

is Hamiltonian with respect to I. Let now

$$z = \begin{pmatrix} u \\ v \end{pmatrix},$$

where $u, v : I \to \mathbb{R}^n$. Then

$$\begin{aligned}
z^\beta &= \begin{pmatrix} u^\beta \\ v^\beta \end{pmatrix}, \\
\mathscr{M}^* \mathscr{M} z^\beta &= \begin{pmatrix} I & O \\ O & O \end{pmatrix} \begin{pmatrix} u^\beta \\ v^\beta \end{pmatrix} \\
&= \begin{pmatrix} u^\beta \\ 0 \end{pmatrix}, \\
\mathscr{M} \mathscr{M}^* z &= \begin{pmatrix} O & O \\ O & I \end{pmatrix} \begin{pmatrix} u \\ v \end{pmatrix} \\
&= \begin{pmatrix} u \\ v \end{pmatrix}, \\
\mathscr{M}^* \mathscr{M} z^\beta + \mathscr{M} \mathscr{M}^* z &= \begin{pmatrix} u^\beta \\ 0 \end{pmatrix} + \begin{pmatrix} 0 \\ v \end{pmatrix} \\
&= \begin{pmatrix} u^\beta \\ v \end{pmatrix}, \\
\mathscr{H} \left(\mathscr{M}^* \mathscr{M} z^\beta + \mathscr{M} \mathscr{M}^* z \right) &= \begin{pmatrix} A & B \\ -C & D \end{pmatrix} \begin{pmatrix} u^\beta \\ v \end{pmatrix} \\
&= \begin{pmatrix} A u^\beta + B v \\ -C u^\beta + D v \end{pmatrix} \\
&= \begin{pmatrix} A u^\beta + B v \\ -C u^\beta - A^* v \end{pmatrix}, \\
D_\beta x &= \begin{pmatrix} D_\beta u \\ D_\beta v \end{pmatrix} \\
&= \begin{pmatrix} A u^\beta + B v \\ -C u^\beta - A^* v \end{pmatrix},
\end{aligned}$$

i.e., we get the system

$$\begin{cases} D_\beta u = Au^\beta + Bv \\ D_\beta v = -Cu^\beta - A^* v. \end{cases}$$

This completes the remark.

4.3 Conjoined Bases

Let $a, b \in I$, $a < \beta^{-1}(b)$. Suppose that A, B and C satisfy the following conditions:

$$A, B, C : [a,b] \to \mathbb{R}^{n\times n}, \quad A, B, C \in \mathscr{C}_\beta([a,b]), \tag{4.3}$$

$$(I - (\beta - I)A)^{-1} \quad exists \quad on \quad [a,b], \tag{4.4}$$

$$B \quad and \quad C \quad are \quad symmetric \quad on \quad [a,b]. \tag{4.5}$$

Note that $(I - (\beta - I)A)^{-1} \in \mathscr{C}_\beta([a,b])$. Consider the system

$$\begin{cases} D_\beta x = Ax^\beta + Bu \\ D_\beta u = -Cx^\beta - A^T u \quad on \quad [a,b]. \end{cases} \tag{4.6}$$

Let

$$\mathscr{H} = \begin{pmatrix} A & B \\ -C & -A^T \end{pmatrix}.$$

Then the system (4.6) we can rewrite in the form

$$D_\beta \begin{pmatrix} x \\ u \end{pmatrix} = \mathscr{H} \left(\mathscr{M}^* \mathscr{M} \begin{pmatrix} x \\ u \end{pmatrix}^\beta + \mathscr{M}\mathscr{M}^* \begin{pmatrix} x \\ u \end{pmatrix} \right).$$

Note that $\mathscr{H}$ is Hamiltonian, and

$$\begin{aligned} I - (\beta - I)\mathscr{H}\mathscr{M}^*\mathscr{M} &= I - (\beta - I) \begin{pmatrix} A & B \\ -C & -A^T \end{pmatrix} \begin{pmatrix} I & O \\ O & O \end{pmatrix} \\ &= I - (\beta - I) \begin{pmatrix} A & O \\ -C & O \end{pmatrix} \\ &= \begin{pmatrix} I - (\beta - I)A & O \\ -C & I \end{pmatrix}. \end{aligned}$$

Hence, using the condition (4.4), we conclude that

$$I - (\beta - I)\mathscr{H}\mathscr{M}^*\mathscr{M}$$

is invertible on $[a,b]$. Therefore, the system (4.6) is a Hamiltonian system.

Definition 4.5 By a solution of (4.6), we mean a pair (x,u) with $x,u \in \mathscr{C}_\beta^1([a,b])$, satisfying the system (4.6) on $[a,b]$.

We use a usual agreement that the vector-valued solutions of (4.6) are denoted by small letters and the $n \times n$ matrix-valued solutions by capital letters.

Theorem 4.7 (Wronskian Identity)

For any two solutions (X,U) and $(\tilde{X},\tilde{U})$ of (4.6) *we have*

$$X^T\tilde{U} - U^T\tilde{X}$$

is a constant on $[a,b]$.

Proof We have

$$\begin{aligned}
D_\beta\left(X^T\tilde{U} - U^t\tilde{X}\right) &= D_\beta\left(X^T\right)\tilde{U} + \left(X^T\right)^\beta D_\beta\tilde{U} \\
&\quad -D_\beta\left(U^T\right)\tilde{X}^\beta - U^T D_\beta\tilde{X} \\
&= \left(D_\beta X\right)^T\tilde{U} + \left(X^\beta\right)^T D_\beta\tilde{U} \\
&\quad - \left(D_\beta U\right)^T\tilde{X}^\beta - U^T D_\beta\tilde{X} \\
&= \left(AX^\beta + BU\right)^T\tilde{U} \\
&\quad + \left(X^\beta\right)^T\left(-C\tilde{X}^\beta - A^T\tilde{U}\right) \\
&\quad - \left(-CX^\beta - A^T U\right)^T\tilde{X}^\beta \\
&\quad -U^T\left(A\tilde{X}^\beta + B\tilde{U}\right) \\
&= \left(X^\beta\right)^T A^T\tilde{U} + U^T B\tilde{U} \\
&\quad - \left(X^\beta\right)^T C\tilde{X}^\beta - \left(X^\beta\right)^T A^T\tilde{U} \\
&\quad + \left(X^\beta\right)^T C\tilde{X}^\beta + U^T A\tilde{X}^\beta \\
&\quad -U^T A\tilde{X}^\beta - U^T B\tilde{U} \\
&= 0 \quad on \quad I.
\end{aligned}$$

This completes the proof. □

Let

$$\tilde{A} = (I - (\beta - I)A)^{-1}$$

and (x,u) is a solution of the system (4.6). Then

$$D_\beta x = Ax^\beta + Bu$$

$$= Ax + (\beta - I)AD_\beta x + Bu,$$

$$D_\beta u = -Cx^\beta - A^T u$$

$$= -Cx - (\beta - I)CD_\beta x - A^T u \quad on \quad I.$$

Hence,

$$(I - (\beta - I)A)D_\beta x = Ax + Bu \quad on \quad I$$

and

$$D_\beta x = \tilde{A}Ax + \tilde{A}Bu \quad on \quad I^\kappa.$$

Then

$$D_\beta u = -Cx - (\beta - I)C(\tilde{A}Ax + \tilde{A}Bu) - A^T u$$

$$= -(C + (\beta - I)C\tilde{A}A)x - ((\beta - I)C\tilde{A}B + A^T)u$$

$$= -C\tilde{A}(I - (\beta - I)A + (\beta - I)A)x - ((\beta - I)C\tilde{A}B + A^T)u$$

$$= -C\tilde{A}x - ((\beta - I)C\tilde{A}B + A^T)u \quad on \quad I.$$

Therefore, the system (4.6) can be rewritten in the form

$$\begin{cases} D_\beta x = \tilde{A}Ax + \tilde{A}Bu \\ D_\beta u = -C\tilde{A}x - ((\beta - I)C\tilde{A}B + A^T)u \quad on \quad I. \end{cases} \tag{4.7}$$

Let

$$w = \begin{pmatrix} x \\ u \end{pmatrix}, \quad S = \begin{pmatrix} \tilde{A}A & \tilde{A}B \\ -C\tilde{A} & -((\beta - I)C\tilde{A}B + A^T) \end{pmatrix}.$$

Then (4.7) takes the form

$$D_\beta w = Sw.$$

We have

$$S^T = \begin{pmatrix} A^T\tilde{A}^T & -\tilde{A}^T C \\ B\tilde{A}^T & -((\beta - I)B\tilde{A}^T C + A) \end{pmatrix},$$

$$S^T J = \begin{pmatrix} A^T\tilde{A}^T & -\tilde{A}^T C \\ B\tilde{A}^T & -((\beta - I)B\tilde{A}^T C + A) \end{pmatrix} \begin{pmatrix} O & I \\ -I & O \end{pmatrix}$$

$$= \begin{pmatrix} \tilde{A}^T C & A^T\tilde{A}^T \\ (\beta - I)B\tilde{A}^T C + A & B\tilde{A}^T \end{pmatrix},$$

$$JS = \begin{pmatrix} O & I \\ -I & O \end{pmatrix} \begin{pmatrix} \tilde{A}A & \tilde{A}B \\ -C\tilde{A} & -((\beta - I)C\tilde{A}B + A^T) \end{pmatrix}$$

$$= \begin{pmatrix} -C\tilde{A} & -((\beta - I)C\tilde{A}B + A^T) \\ -\tilde{A}A & -\tilde{A}B \end{pmatrix},$$

$$S^T J + JS = \begin{pmatrix} \tilde{A}^T C & A^T\tilde{A}^T \\ (\beta - I)B\tilde{A}^T C + A & B\tilde{A}^T \end{pmatrix} + \begin{pmatrix} -C\tilde{A} & -((\beta - I)C\tilde{A}B + A^T) \\ -\tilde{A}A & -\tilde{A}B \end{pmatrix}$$

$$= \begin{pmatrix} \tilde{A}^T C - C\tilde{A} & A^T\tilde{A}^T - ((\beta - I)C\tilde{A}B + A^T) \\ (\beta - I)B\tilde{A}^T C + A - \tilde{A}A & B\tilde{A}^T - \tilde{A}B \end{pmatrix},$$

$$S^T JS = \begin{pmatrix} \tilde{A}^T C & A^T\tilde{A}^T \\ (\beta - I)B\tilde{A}^T C + A & B\tilde{A}^T \end{pmatrix} \begin{pmatrix} \tilde{A}A & \tilde{A}B \\ -C\tilde{A} & -((\beta - I)C\tilde{A}B + A^T) \end{pmatrix}$$

$$= \begin{pmatrix} \tilde{A}^T C\tilde{A}A - A^T\tilde{A}^T C\tilde{A} & \tilde{A}^T C\tilde{A}B - A^T\tilde{A}^T((\beta - I)C\tilde{A}B + A^T) \\ (\beta - I)B\tilde{A}^T C\tilde{A}A + A\tilde{A}A - B\tilde{A}^T C\tilde{A} & A\tilde{A}B - B\tilde{A}^T A^T \end{pmatrix}.$$

Note that

$$\tilde{A}^T C - C\tilde{A} + (\beta - I)\tilde{A}^T C\tilde{A}A - (\beta - I)A^T\tilde{A}^T C\tilde{A}$$

$$= \tilde{A}^T C(I + (\beta - I)\tilde{A}A) - (I + (\beta - I)A^T\tilde{A}^T)C\tilde{A}$$

$$= \tilde{A}^T C\tilde{A}(I - (\beta - I)A + (\beta - I)A) - (I - (\beta - I)A^T + (\beta - I)A^T)\tilde{A}^T C\tilde{A}$$

$$= 0,$$

$$A^T\tilde{A}^T - ((\beta - I)C\tilde{A}B + A^T) + (\beta - I)\tilde{A}^T C\tilde{A}B - (\beta - I)^2 A^T\tilde{A}^T C\tilde{A}B$$

$$-(\beta - I)A^T\tilde{A}^T A^T$$

$$= A^T\tilde{A}^T(I - (\beta - I)A^T) - (\beta - I)(I + (\beta - I)A^T\tilde{A}^T)(\tilde{A}B - A^T + (\beta - I)\tilde{A}^T C\tilde{A}B)$$

$$= A^T - (\beta - I)(I - (\beta - I)A^T + (\beta - I)A^T)\tilde{A}^T C\tilde{A}B - A^T - (\beta - I)\tilde{A}^T C\tilde{A}B$$

$$= 0,$$

$$(\beta - I)B\tilde{A}^T C + A - \tilde{A}A + (\beta - I)^2 B\tilde{A}^T C\tilde{A}A + (\beta - I)A\tilde{A}A - (\beta - I)B\tilde{A}^T C\tilde{A}$$

$$= (\beta - I)B\tilde{A}^T C\tilde{A}(I - (\beta - I)A + (\beta - I)A) - (\beta - I)B\tilde{A}^T C\tilde{A} + A - A$$

$$= 0,$$

$$B\tilde{A}^T - \tilde{A}B + (\beta - I)A\tilde{A}B - (\beta - I)B\tilde{A}^T A^T$$

$$= B\tilde{A}^T (I - (\beta - I)A^T) - (I - (\beta - I)A)\tilde{A}B$$

$$= B - B$$

$$= 0.$$

Therefore,

$$S^T J + JS + (\beta - I)S^T JS = O$$

and the system (4.7) is symplectic. Since S is symplectic with respect to I, by Theorem 4.4, it follows that it is regressive. From this and, we conclude that the IVP for the system (4.7) with initial condition

$$x(s) = \overline{x}, \quad u(s) = \overline{u},$$

$s \in [a,b]$, $\overline{x}, \overline{u} \in \mathbb{R}^n$, has unique solution on $[a,b]$. For a solution (X,U) of (4.6), we have

$$X^\beta = X + (\beta - I)D_\beta X$$

$$= X + (\beta - I)AX^\beta + (\beta - I)BU \quad on \quad [a,b],$$

whereupon

$$(I - (\beta - I)A)X^\beta = X + (\beta - I)BU \quad on \quad [a,b],$$

or

$$X^\beta = \tilde{A}X + (\beta - I)\tilde{A}BU \quad on \quad [a,b],$$

and

$$\begin{aligned} U^{\beta} &= U + (\beta - I)D_{\beta}U \\ &= U - (\beta - I)CX^{\beta} - (\beta - I)A^{T}U \\ &= -(\beta - I)CX^{\beta} + (I - (\beta - I)A^{T})U \\ &= -(\beta - I)C(\tilde{A}X + (\beta - I)\tilde{A}BU) + (\tilde{A}^{T})^{-1}U \\ &= -(\beta - I)C\tilde{A}X + (-(\beta - I)^{2}C\tilde{A}B + (\tilde{A}^{T})^{-1})U \quad on[a,b], \end{aligned}$$

i.e., we get the system

$$\begin{cases} X^{\beta} = \tilde{A}X + (\beta - I)\tilde{A}BU \\ U^{\beta} = -(\beta - I)C\tilde{A}X + (-(\beta - I)^{2}C\tilde{A}B + (\tilde{A}^{T})^{-1})U \quad on \quad [a,b]. \end{cases}$$

From the last system, we obtain

$$(\beta - I)CX^{\beta} + U^{\beta} = (\tilde{A}^{T})^{-1}U \quad on \quad [a,b],$$

or

$$U = (\beta - I)\tilde{A}^{T}CX^{\beta} + \tilde{A}^{T}U^{\beta} \quad on \quad [a,b],$$

and

$$\tilde{A}X = X^{\beta} - (\beta - I)^{2}\tilde{A}B\tilde{A}^{T}CX^{\beta} - (\beta - I)\tilde{A}B\tilde{A}^{T}U^{\beta} \quad on \quad [a,b],$$

or

$$X = (\tilde{A}^{-1} - (\beta - I)^{2}B\tilde{A}^{T}C)X^{\beta} - (\beta - I)B\tilde{A}^{T}U^{\beta} \quad on \quad [a,b].$$

In this way we obtain the system

$$\begin{cases} X = (\tilde{A}^{-1} - \mu^{2}B\tilde{A}^{T}C)X^{\beta} - (\beta - I)B\tilde{A}^{T}U^{\beta} \\ U = (\beta - I)\tilde{A}^{T}CX^{\beta} + \tilde{A}^{T}U^{\beta} \quad on \quad [a,b]. \end{cases}$$

Definition 4.6 A solution (X,U) of the system (4.6) is called a basis, if

$$rank(X^{T}U^{T}) = n$$

at some, and hence at any, $t \in [a,b]$.

Definition 4.7 A solution (X,U) of (4.6) is called a conjoined solution, if

$$X^{T}U - U^{T}X = O \quad on \quad [a,b].$$

Definition 4.8 Two conjoined bases (X,U), $(\tilde{X},\tilde{U})$ are called normalized, if

$$X^T\tilde{U} - U^T\tilde{X} = I \quad on \quad [a,b].$$

Definition 4.9 The unique solution of the IVP for the system (4.6) with initial condition

$$X(a) = O, \quad U(a) = I,$$

is called the principal solution at a.

Definition 4.10 The unique solution (X,U) of the system (4.6) with initial condition

$$X(a) = -I, \quad U(a) = O,$$

is called the associated solution at a.

Definition 4.11 Together, the principal solution at a and the associated solution at a of the system (4.6), is called the special normalized bases of (4.6) at a.

Remark 4.2 The principal and the associated solution of (4.6) at a form normalized conjoined bases of (4.6).

Theorem 4.8 *Let* (X,U), $(\tilde{X},\tilde{U})$ *be any solutions of the system* (4.6). *Then they form normalized conjoined bases of* (4.6) *if and only if the matrix*

$$\begin{pmatrix} X & \tilde{X} \\ U & \tilde{U} \end{pmatrix}$$

is symplectic. In this case the following identities hold on $[a,b]^\kappa$.

1. X^TU, $\tilde{X}^T\tilde{U}$, $X\tilde{X}^T$, $U\tilde{U}^T$ *are symmetric,*
2. $$X\tilde{U}^T - \tilde{X}U^T = \tilde{U}X^T - U\tilde{X}^T = I,$$
3. $$X^\beta\tilde{U}^T - \tilde{X}^\beta U^T = \tilde{A},$$
4. $$\tilde{X}^\beta X^T - X^\beta\tilde{X}^T = (\beta - I)\tilde{A}B,$$

5.
$$\tilde{U}^{\beta}U^T - U^{\beta}\tilde{U}^T = (\beta - I)C\tilde{A},$$

6.
$$U^{\beta}\tilde{X}^T - \tilde{U}^{\beta}X^T = (\beta - I)^2 C\tilde{A}B - \left(\tilde{A}^T\right)^{-1}.$$

Proof Let (X,U), $(\tilde{X},\tilde{U})$ form a normalized conjoined bases of the system (4.6). Then

$$X^TU - U^TX = O, \quad \tilde{X}^T\tilde{U} - \tilde{U}^T\tilde{X} = O \quad on \quad [a,b], \tag{4.8}$$

$$X^T\tilde{U} - U^T\tilde{X} = I \quad on \quad [a,b]. \tag{4.9}$$

Let

$$S = \begin{pmatrix} X & \tilde{X} \\ U & \tilde{U} \end{pmatrix}.$$

Then

$$S^T = \begin{pmatrix} X^T & U^T \\ \tilde{X}^T & \tilde{U}^T \end{pmatrix}$$

and

$$\begin{aligned} S^T \mathscr{I} S &= \begin{pmatrix} X^T & U^T \\ \tilde{X}^T & \tilde{U}^T \end{pmatrix} \begin{pmatrix} O & I \\ -I & O \end{pmatrix} \begin{pmatrix} X & \tilde{X} \\ U & \tilde{U} \end{pmatrix} \\ &= \begin{pmatrix} X^T & U^T \\ \tilde{X}^T & \tilde{U}^T \end{pmatrix} \begin{pmatrix} U & \tilde{U} \\ -X & -\tilde{X} \end{pmatrix} \\ &= \begin{pmatrix} X^TU - U^TX & X^T\tilde{U} - U^T\tilde{X} \\ \tilde{X}^TU - \tilde{U}^TX & \tilde{X}^T\tilde{U} - \tilde{U}^T\tilde{X} \end{pmatrix} \qquad (4.10) \\ &= \begin{pmatrix} O & I \\ -I & O \end{pmatrix} \\ &= \mathscr{I}, \end{aligned}$$

where we have used (4.8) and (4.9). Therefore, S is symplectic. If we suppose that S is symplectic, by (4.10), we get (4.8) and (4.9).

1. By (4.8), it follows that the matrices X^TU and $\tilde{X}^T\tilde{U}$ are symmetric matrices. Since the matrix S is symplectic, by Exercise 4.1, it follows that the matrices $X\tilde{X}^T$ and $U\tilde{U}^T$ are symmetric.
2. Again by Exercise 4.1, we obtain that

   $$X\tilde{U}^T - \tilde{X}U^T = I,$$

   whereupon

   $$\tilde{U}X^T - U^T\tilde{X} = I.$$

3. We have, using 1 and 2,

$$\begin{aligned} X^{\beta}\tilde{U}^T - \tilde{X}^{\beta}U^T &= \left(\tilde{A}X + (\beta - I)\tilde{A}BU\right)\tilde{U}^T \\ &\quad - \left(\tilde{A}\tilde{X} + (\beta - I)\tilde{A}B\tilde{U}\right)U^T \\ &= \tilde{A}X\tilde{U}^T + (\beta - I)\tilde{A}BU\tilde{U}^T \\ &\quad -\tilde{A}\tilde{X}U^T - (\beta - I)\tilde{A}B\tilde{U}U^T \\ &= \tilde{A}(X\tilde{U}^T - \tilde{X}U^T) + (\beta - I)\tilde{A}B(U\tilde{U}^T - \tilde{U}U^T) \\ &= \tilde{A}. \end{aligned}$$

4. By 2, we have

$$\tilde{U}X^T - U\tilde{X}^T = I$$

and

$$\begin{aligned} \tilde{X}^{\beta}X^T - X^{\beta}\tilde{X}^T &= (\tilde{A}\tilde{X} + (\beta - I)\tilde{A}B\tilde{U})X^T \\ &\quad -(\tilde{A}X + (\beta - I)\tilde{A}BU)\tilde{X}^T \\ &= \tilde{A}\tilde{X}X^T + (\beta - I)\tilde{A}B\tilde{U}X^T \\ &\quad -\tilde{A}X\tilde{X}^T - (\beta - I)\tilde{A}BU\tilde{X}^T \\ &= (\beta - I)\tilde{A}B(\tilde{U}X^T - U\tilde{X}^T) \\ &= (\beta - I)\tilde{A}B. \end{aligned}$$

5. We have

$$\begin{aligned} \tilde{U}^{\beta}U^T - U^{\beta}\tilde{U}^T &= -(\beta - I)C\tilde{A}\tilde{X}U^T + (-(\beta - I)^2C\tilde{A}B + (\tilde{A}^T)^{-1})\tilde{U}U^T \\ &\quad +(\beta - I)C\tilde{A}X\tilde{U}^T + (-(\beta - I)^2C\tilde{A}B + (\tilde{A}^T)^{-1})U\tilde{U}^T \\ &= (\beta - I)C\tilde{A}(X\tilde{U}^T - \tilde{X}U^T) \\ &= (\beta - I)C\tilde{A}. \end{aligned}$$

6. We have

$$U^{\beta}\tilde{X}^{T}-\tilde{U}^{\beta}XT=-(\beta-I)C\tilde{A}X\tilde{X}^{T}+\left(-(\beta-I)^{2}C\tilde{A}B+\left(\tilde{A}^{T}\right)^{-1}\right)U\tilde{X}^{T}$$
$$+(\beta-I)C\tilde{A}\tilde{X}X^{T}-\left(-(\beta-I)^{2}C\tilde{A}B+\left(\tilde{A}^{T}\right)^{-1}\right)\tilde{U}X^{T}$$
$$=\left((\beta-I)^{2}C\tilde{A}B-\left(\tilde{A}^{T}\right)^{-1}\right)\left(\tilde{U}X^{T}-U\tilde{X}^{T}\right)$$
$$=(\beta-I)^{2}C\tilde{A}B-\left(\tilde{A}^{T}\right)^{-1}.$$

This completes the proof. □

Theorem 4.9 *Let (X,U) be a conjoined basis of* (4.6)*. Then there exists another conjoined basis $(\tilde{X},\tilde{U})$ of* (4.6) *such that (X,U) and $(\tilde{X},\tilde{U})$ are normalized.*

Proof Let $t_0\in[a,b]$. Since (X,U) is a conjoined basis of (4.6) we have that

$$(X^{T}X+U^{T}U)^{-1}(t_0)$$

exists. Let $(\tilde{X},\tilde{U})$ be the solution of equation (4.6) with initial condition

$$\begin{pmatrix}\tilde{X}(t_0)\\ \tilde{U}(t_0)\end{pmatrix}=\begin{pmatrix}-U(t_0)(X^{T}X+U^{T}U)^{-1}(t_0)\\ X(t_0)(X^{T}X+U^{T}U)^{-1}(t_0)\end{pmatrix}.$$

Then

$$\tilde{X}^{T}(t_0)\tilde{U}(t_0)-\tilde{U}^{T}(t_0)\tilde{X}(t_0)$$
$$=-\left(\left(X^{T}X+U^{T}U\right)^{-1}\right)^{T}(t_0)U^{T}(t_0)X(t_0)\left(X^{T}X+U^{T}U\right)^{-1}(t_0)$$
$$+\left(\left(X^{T}X+U^{T}U\right)^{-1}\right)^{T}(t_0)X^{T}(t_0)U(t_0)\left(X^{T}X+U^{T}U\right)^{-1}(t_0)$$
$$=\left(\left(X^{T}X+U^{T}U\right)^{-1}\right)^{T}(t_0)\left(X^{T}(t_0)U(t_0)-U^{T}(t_0)X(t_0)\right)$$
$$\times(X^{T}X+U^{T}U)^{-1}(t_0)$$
$$=O.$$

Therefore, $(\tilde{X},\tilde{U})$ is a conjoined basis of (4.6). From this and, we get

$$X^{T}\tilde{U}-U^{T}\tilde{X}$$

is a constant and this constant is

$$
\begin{aligned}
&X^T(t_0)\tilde{U}(t_0) - U^T(t_0)\tilde{X}(t_0)\\
&= X^T(t_0)X(t_0)\left(X^TX + U^TU\right)^{-1}(t_0) + U^T(t_0)U(t_0)\left(X^TX + U^TU\right)^{-1}(t_0)\\
&= \left(X^TX + U^TU\right)(t_0)\left(X^TX + U^TU\right)^{-1}(t_0)\\
&= I.
\end{aligned}
$$

This completes the proof. □

Theorem 4.10 *Let $t \in [a,b]$ and (X,U) be a conjoined solution of* (4.6) *with*

$$
KerX^{\beta}(t) \subset KerX(t).
$$

Then

$$
Ker(X^{\beta})^T \subseteq Ker((\beta - I)B\tilde{A}^T)(t).
$$

Proof By Theorem 4.9, it follows that there exists another conjoined solution $(\tilde{X}, \tilde{U})$ of (4.6) such that (X,U) and $(\tilde{X}, \tilde{U})$ are normalized. Let $\alpha \in Ker(X^{\beta})^T(t)$. Then

$$
(X^{\beta})^T(t)\alpha = 0.
$$

Hence, using Theorem 4.8, we get

$$
\begin{aligned}
X^{\beta}(t)\left(\tilde{X}^{\beta}\right)^T(t)\alpha &= \left(\tilde{X}^{\beta}\right)(t)\left(X^{\beta}\right)^T(t)\alpha\\
&= 0.
\end{aligned}
$$

Therefore,

$$
\begin{aligned}
Ker\left(\tilde{X}^{\beta}\right)^T &\subseteq Ker(X^{\beta})(t)\\
&\subset KerX(t).
\end{aligned}
$$

Consequently,

$$
X(t)\left(\tilde{X}^{\beta}\right)^T(t)\alpha = 0.
$$

By Theorem 4.8, item 4, we obtain

$$
((\beta - I)\tilde{A}B)^T(t) = X\left(\tilde{X}^{\beta}\right)^T(t) - \tilde{X}(X^{\beta})^T(t).
$$

Then

$$((\beta - I)\tilde{A}B)^T(t)\alpha = X(\tilde{X}^\beta)^T(t)\alpha - \tilde{X}(X^\beta)^T(t)\alpha$$

$$= 0.$$

Consequently, $\alpha \in Ker((\beta - I)\tilde{A}B)^T$. This completes the proof. □

Definition 4.12 By $M^\dagger$ we denote the Moore-Penrose generalized inverse of the matrix M, i.e., the unique matrix satisfying

1. $MM^\dagger M = M$,
2. $M^\dagger MM^\dagger = M^\dagger$,
3. $MM^\dagger$ and $M^\dagger M$ are symmetric.

Definition 4.13 A conjoined basis (X, U) of the system (4.6) is said to have no focal point in $(a, b]$, provided that X is invertible for all dense points $t \in (a, b]$ and

$$KerX^\beta \subseteq KerX, \quad D = X(X^\beta)^T\tilde{A}B \geq 0 \quad on \quad [a, b]^\kappa.$$

Definition 4.14 The system (4.6) is called disconjugate on $[a, b]$, if the principal solution of (4.6) at a has no focal point in $(a, b]$.

4.4 Riccati Equations

In this section we will use the notations from the previous sections of this chapter. We consider the Riccati operator

$$R(Q) = D_\beta Q + C + A^T Q + (Q^\beta + (\beta - I)C)\tilde{A}(A + BQ) \quad on \quad [a, b].$$

We have

$$(\beta - I)R(Q) = (\beta - I)D_\beta Q + (\beta - I)C + (\beta - I)A^T Q$$

$$+ (\beta - I)(Q^\beta + (\beta - I)C)\tilde{A}(A + BQ)$$

$$= Q^\beta - Q + (\beta - I)C + (\beta - I)A^T Q + (\beta - I)(Q^\beta$$

$$+ (\beta - I)C)\tilde{A}(A + BQ)$$

$$= Q^\beta - Q + (\beta - I)C + (\beta - I)A^T Q + (Q^\beta$$

$$+(\beta - I)C)\tilde{A}((\beta - I)A + (\beta - I)BQ)$$

$$= Q^\beta - Q + (\beta - I)C + (\beta - I)A^T Q$$

$$+(Q^\beta + (\beta - I)C)\tilde{A}((\beta - I)A - I + I + (\beta - I)BQ)$$

$$= Q^\beta - Q + (\beta - I)C + (\beta - I)A^T Q - (Q^\beta + (\beta - I)C)\tilde{A}\tilde{A}^{-1}$$

$$+(Q^\beta + (\beta - I)C)\tilde{A}(I + (\beta - I)BQ)$$

$$= (\beta - I)A^T Q + (Q^\beta + (\beta - I)C)\tilde{A}(I + (\beta - I)BQ) - Q$$

$$= -(I - (\beta - I)A^T)Q + (Q^\beta + (\beta - I)C)\tilde{A}(I + (\beta - I)BQ)$$

$$= -(\tilde{A}^T)^{-1}Q + (Q^\beta + (\beta - I)C)\tilde{A}(I + (\beta - I)BQ) \quad on \quad [a,b],$$

whereupon

$$(\beta - I)\tilde{A}^T R(Q) = \tilde{A}^T(Q^\beta + (\beta - I)C)\tilde{A}(I + (\beta - I)BQ) - Q, \quad on \quad [a,b].$$

The next result gives a criterion for existence of a symmetric solution of the Riccati equation.

Theorem 4.11 *The Riccati matrix equation*

$$R(Q) = D_\beta Q + C + A^T Q + (Q^\beta + (\beta - I)C)\tilde{A}(A + BQ) = 0 \tag{4.11}$$

has a symmetric solution Q on $[a,b]$ with $I + (\beta - I)BQ$ nonsingular and $(I + \mu BQ)^{-1}B \geq 0$ on $[a,b]$ if and only if the system (4.6) has a conjoined basis (X,U) such that X is invertible on $[a,b]$ with $D = X\left(X^\beta\right)^{-1}\tilde{A}B \geq 0$ on $[a,b]$. Moreover, in this case, $Q = UX^{-1}$ on $[a,b]$ and

$$D = (I + (\beta - I)BQ)^{-1}B \quad on \quad [a,b].$$

Proof 1. Let (4.6) have a conjoined basis on $[a,b]$ with X invertible on $[a,b]$. Let $Q = -UX^{-1}$ on $[a,b]$. We have that Q is symmetric and on $[a,b]$

$$
\begin{aligned}
D_\beta Q &= D_\beta(UX^{-1})\\
&= UD_\beta(X^{-1}) + D_\beta U(X^{-1})^\beta\\
&= -UX^{-1}D_\beta X(X^\beta)^{-1} + D_\beta U(X^\beta)^{-1}\\
&= (-UX^{-1}D_\beta X + D_\beta U)(X^\beta)^{-1}\\
&= (-QD_\beta X + D_\beta U)(X^{-1})^\beta\\
&= \left(-Q(AX^\beta + BU) - CX^\beta - A^T U\right)(X^{-1})^\beta\\
&= \left(-QAX^\beta - QBU - CX^\beta - A^T U\right)(X^{-1})^\beta\\
&= -QA - C - (QB + A^T)U(X^{-1})^\beta\\
&= -QA - C - (QB + A^T)((\beta - I)\tilde{A}^T CX^\beta + \tilde{A}^T U^\beta)(X^{-1})^\beta\\
&= -QA - C - (QB + A^T)((\beta - I)\tilde{A}^T C + \tilde{A}^T (UX^{-1})^\beta)\\
&= -QA - C - (QB + A^T)((\beta - I)\tilde{A}^T C + \tilde{A}^T Q^\beta)\\
&= -QA - C - (QB + A^T)\tilde{A}^T((\beta - I)C + Q^\beta),
\end{aligned}
$$

where we have used

$$U = (\beta - I)\tilde{A}^T CX^\beta + \tilde{A}^T U^\beta \quad on \quad [a,b].$$

Next,

$$
\begin{aligned}
I + (\beta - I)BQ &= I + (\beta - I)BUX^{-1}\\
&= XX^{-1} + (\beta - I)BUX^{-1}\\
&= (X + (\beta - I)BU)X^{-1} \quad on \quad [a,b].
\end{aligned}
$$

Since

$$X^{\beta} = \tilde{A}X + (\beta - I)\tilde{A}BU \quad on \quad [a,b],$$

we get

$$X^{\beta} = \tilde{A}(X + (\beta - I)BU) \quad on \quad [a,b],$$

and

$$X + (\beta - I)BU = \tilde{A}^{-1}X^{\beta} \quad on \quad [a,b].$$

Therefore,

$$I + (\beta - I)BQ\tilde{A}^{-1} = X^{\beta}X^{-1} \quad on \quad [a,b].$$

Hence,

$$\begin{aligned}(I + (\beta - I)BQ)^{-1}B &= (\tilde{A}^{-1}X^{\beta}X^{-1})^{-1}B \\ &= X(X^{\beta})^{-1}\tilde{A}B \\ &= D \\ &\geq 0 \quad on \quad [a,b].\end{aligned}$$

2. Let Q be a symmetric solution of (4.11) on $[a,b]$ with $I + (\beta - I)BQ$ nonsingular and $(I + (\beta - I)BQ)^{-1}B \geq 0$ on $[a,b]$. Note that

$$\begin{aligned}I + (\beta - I)\tilde{A}(A + BQ) &= \tilde{A}(\tilde{A}^{-1} + (\beta - I)A + (\beta - I)BQ) \\ &= \tilde{A}(I - (\beta - I)A + (\beta - I)A + (\beta - I)BQ) \\ &= \tilde{A}(I + (\beta - I)BQ) \quad on \quad [a,b].\end{aligned}$$

Therefore, the matrix $\tilde{A}(A + BQ)$ is regressive on $[a,b]$. Consequently, the IVP

$$D_{\beta}X = \tilde{A}(A + BQ)X, \quad X(s) = I,$$

for some $s \in [a,b]$, has a unique solution X and it is nonsingular. We set

$$U = QX.$$

Then

$$
\begin{aligned}
D_\beta X &= \tilde{A}(A+BQ)X \\
&= \tilde{A}(AX+BQX) \\
&= \tilde{A}(AX+BU) \\
&= \tilde{A}(A(X^\beta-(\beta-I)D_\beta X)+BU) \\
&= \tilde{A}((AX^\beta+BU)-(\beta-I)AD_\beta X) \\
&= \tilde{A}(AX^\beta+BU)-(\beta-I)\tilde{A}AD_\beta X \\
&= \tilde{A}(AX^\beta+BU)+\tilde{A}(I-(\beta-I)A-I)D_\beta X \\
&= \tilde{A}(AX^\beta+BU)+\tilde{A}(I-(\beta-I)A)D_\beta X-\tilde{A}D_\beta X \\
&= \tilde{A}(AX^\beta+BU-D_\beta X)+D_\beta X \quad on \quad [a,b],
\end{aligned}
$$

whereupon

$$
D_\beta X = AX^\beta+BU \quad on \quad [a,b]. \tag{4.12}
$$

Next,

$$
\begin{aligned}
D_\beta U &= D_\beta(QX) \\
&= Q^\beta D_\beta X + D_\beta QX \\
&= Q^\beta D_\beta X + (-C-A^TQ-(Q^\beta+(\beta-I)C)\tilde{A}(A+BQ))X \\
&= Q^\beta D_\beta X - CX - A^TQX - (Q^\beta+(\beta-I)C)\tilde{A}(A+BQ)X \\
&= Q^\beta D_\beta X - CX - A^TU - (Q^\beta+(\beta-I)C)D_\beta X \\
&= -CX - A^TU - (\beta-I)CD_\beta X \\
&= -C(X+(\beta-I)D_\beta X) - A^TU \\
&= -CX^\beta - A^TU \quad on \quad [a,b].
\end{aligned}
$$

Also, we have that

$$rank(X^T U^T) = n$$

and

$$X^T U = X^T QX, \quad U^T X = X^T QX,$$

whereupon

$$X^T U - U^T X = 0.$$

Therefore, (X,U) is a conjoined basis of the system (4.6) with X invertible on $[a,b]$. By (4.12), we obtain

$$(\beta - I)D_\beta X = (\beta - I)AX^\beta + (\beta - I)BU \quad on \quad [a,b],$$

or

$$X^\beta - X = (\beta - I)AX^\beta + (\beta - I)BU \quad on \quad [a,b],$$

or

$$(I - (\beta - I)A)X^\beta = X + (\beta - I)BU \quad on \quad [a,b],$$

or

$$\begin{aligned} X^\beta &= \tilde{A}(X + (\beta - I)BU) \\ &= \tilde{A}(X + (\beta - I)BUX^{-1}X) \\ &= \tilde{A}(I + (\beta - I)BQ)X \quad on \quad [a,b]. \end{aligned}$$

Therefore,

$$(X^\beta)^{-1} = X^{-1}(I + (\beta - I)BQ)^{-1}\tilde{A}^{-1} \quad on \quad [a,b],$$

and

$$\begin{aligned} D &= X(X^\beta)^{-1}\tilde{A}B \\ &= XX^{-1}(I + (\beta - I)BQ)^{-1}\tilde{A}^{-1}\tilde{A}B \\ &= (I + (\beta - I)BQ)^{-1}B \\ &\geq 0 \quad on \quad [a,b]^\kappa. \end{aligned}$$

This completes the proof. □

4.5 The Picone Identity

Definition 4.15 The equation

$$D_\beta x = Ax^\beta + Bu$$

is referred to as an equation of motion, while we call

$$D_\beta u = -Cx^\beta - A^T u$$

the Euler equation.

Definition 4.16 (x,u) is called admissible if it satisfies the equation of motion.

Lemma 4.1 *Let (x,u) be admissible, Q is symmetric on $[a,b]$ and differentiable on $[a,b]^\kappa$. We set*

$$z = u - Qx,$$

$$D = B - (\beta - I)B\tilde{A}^T(Q^\beta + (\beta - I)C)\tilde{A}B \quad on \quad [a,b].$$

Then

$$\begin{aligned} & D_\beta(x^T QX) + (x^\beta)^T Cx^\beta - u^T Bu + z^T Dz \\ & = x^T(\tilde{A}^T R(Q) - (\beta - I)QB\tilde{A}^T R(Q))x \\ & \quad +2u^T((\beta - I)B\tilde{A}^T R(Q))x \quad on \quad [a,b] \end{aligned} \tag{4.13}$$

and

$$x + (\beta - I)Dz - (\beta - I)^2 B\tilde{A}^T R(Q)x = (\tilde{A}^{-1} - (\beta - I)B\tilde{A}^T(Q^\beta + (\beta - I)C))x^\beta \tag{4.14}$$

on $[a,b]$.

Proof Firstly, we will prove (4.13). We have

$$D_\beta(x^T Qx) + (x^\beta)^T Cx^\beta - u^T Bu + z^T Dz$$

$$= (x^\beta)^T(D_\beta Qx^\beta + QD_\beta x) + (D_\beta x)^T Qx + (x^\beta)^T Cx^\beta - u^T Bu + z^T Dz$$

$$= (x^\beta)^T D_\beta Qx^\beta + (x^\beta)^T QD_\beta x + (D_\beta x)^T Qx + (x^\beta)^T Cx^\beta - u^T Bu + z^T Dz$$

$$
\begin{aligned}
&= \left(x^T\tilde{A}^T + (\beta - I)u^T B\tilde{A}^T\right) D_\beta Q(\tilde{A}x + (\beta - I)\tilde{A}Bu) \\
&\quad + \left(x^T\tilde{A}^T + (\beta - I)u^T B\tilde{A}^T\right) Q\tilde{A}(Ax + Bu) \\
&\quad + \left(x^T A^T + u^T B\right)\tilde{A}^T Qx \\
&\quad + \left(x^T\tilde{A}^T + (\beta - I)u^T B\tilde{A}^T\right) C(\tilde{A}x + (\beta - I)\tilde{A}Bu) - u^T Bu \\
&\quad + (u^T - x^T Q)D(u - Qx) \\
&= x^T\left(\tilde{A}^T D_\beta Q\tilde{A} + \tilde{A}^T Q\tilde{A}A + A^T\tilde{A}^T Q + \tilde{A}^T C\tilde{A} + QDQ\right)x \\
&\quad + x^T\left((\beta - I)\tilde{A}^T D_\beta Q\tilde{A}B + \tilde{A}^T Q\tilde{A}B + (\beta - I)\tilde{A}^T C\tilde{A}B - QD\right)u \\
&\quad + u^T\left((\beta - I)B\tilde{A}^T D_\beta Q\tilde{A} + (\beta - I)B\tilde{A}^T Q\tilde{A}A + B\tilde{A}^T Q + (\beta - I)B\tilde{A}^T C\tilde{A} - DQ\right)x \\
&\quad + u^T\left((\beta - I)^2 B\tilde{A}^T D_\beta Q\tilde{A}B + (\beta - I)B\tilde{A}^T Q\tilde{A}B + (\beta - I)^2 B\tilde{A}^T C\tilde{A}B - B + D\right)u,
\end{aligned}
$$

on $[a,b]$. Note that

$$
\begin{aligned}
\tilde{A} &= I + (\beta - I)\tilde{A}A \\
&= I + (\beta - I)A\tilde{A}, \\
\mu A\tilde{A} &= (\beta - I)\tilde{A}A \\
&= \tilde{A} - I, \quad on \quad [a,b].
\end{aligned}
$$

Hence, we get

$$
\begin{aligned}
&x^T\left(\tilde{A}^T D_\beta Q\tilde{A} + \tilde{A}^T Q\tilde{A}A + A^T\tilde{A}^T Q + \tilde{A}^T C\tilde{A} + QDQ\right)x \\
&= x^T\Bigg(\tilde{A}^T D_\beta Q + (\beta - I)\tilde{A}^T D_\beta Q\tilde{A}A + \tilde{A}^T Q\tilde{A}A + A^T\tilde{A}^T Q + \tilde{A}^T C \\
&\quad + (\beta - I)\tilde{A}^T CA\tilde{A} + QBQ - (\beta - I)QB\tilde{A}^T(Q^\beta + (\beta - I)C)\tilde{A}BQ\Bigg)x
\end{aligned}
$$

$$
\begin{aligned}
&= x^T \Bigg(\tilde{A}^T D_\beta Q + \tilde{A}^T C + \tilde{A}^T A^T Q + \tilde{A}^T Q\tilde{A}A + (\beta - I)\tilde{A}^T CA\tilde{A} \\
&\quad + \tilde{A}^T Q^\beta \tilde{A}A - \tilde{A}^T Q\tilde{A}A \\
&\quad + QBQ - \mu QB\tilde{A}^T (Q^\beta + (\beta - I)C)\tilde{A}BQ \Bigg) x \\
&= x^T \Bigg(\tilde{A}^T Q^{D_\beta} + \tilde{A}^T C + \tilde{A}^T A^T Q + \tilde{A}^T (Q^\beta + (\beta - I)C)A\tilde{A} \\
&\quad + QBQ - (\beta - I)QB\tilde{A}^T (Q^\beta + (\beta - I)C)\tilde{A}BQ \Bigg) x \\
&= x^T \Bigg(\tilde{A}^T D_\beta Q + \tilde{A}^T C + \tilde{A}^T A^T Q + \tilde{A}^T (Q^\beta + (\beta - I)C)\tilde{A}(A + BQ) \\
&\quad - \tilde{A}^T (Q^\beta + (\beta - I)C)\tilde{A}BQ + QBQ - (\beta - I)QB\tilde{A}^T (Q^\beta + (\beta - I)C)\tilde{A}BQ \Bigg) x \\
&= x^T \Bigg(\tilde{A}^T R(Q) - (\beta - I)QB\tilde{A}^T (Q^\beta + (\beta - I)C)\tilde{A}BQ \\
&\quad - (\beta - I)QB\tilde{A}^T (Q^\beta + (\beta - I)C)\tilde{A}A + (\beta - I)QB\tilde{A}^T (Q^\beta + (\beta - I)C)\tilde{A}A \\
&\quad - \tilde{A}^T (Q^\beta + (\beta - I)C)\tilde{A}BQ + QBQ \Bigg) x \\
&= x^T \Bigg(\tilde{A}^T R(Q) - (\beta - I)QB\tilde{A}^T (Q^\beta + (\beta - I)C)\tilde{A}(A + BQ) \\
&\quad - (\beta - I)QB\tilde{A}^T D_\beta Q - (\beta - I)QB\tilde{A}^T C - (\beta - I)QB\tilde{A}^T A^T Q \\
&\quad + (\beta - I)QB\tilde{A}^T D_\beta Q + (\beta - I)QB\tilde{A}^T C + (\beta - I)QB\tilde{A}^T A^T Q \\
&\quad + (\beta - I)QB\tilde{A}^T (Q^\beta + (\beta - I)C)\tilde{A}A \\
&\quad - \tilde{A}^T (Q^\beta + (\beta - I)C)\tilde{A}BQ + QBQ \Bigg) x
\end{aligned}
$$

$$
\begin{aligned}
&= x^T \Bigg(\tilde{A}^T R(Q) - (\beta - I) QB\tilde{A}^T R(Q) \\
&\quad + QB\tilde{A}^T (Q^\beta - Q) + (\beta - I) QB\tilde{A}^T C + (\beta - I) QB\tilde{A}^T A^T Q \\
&\quad + (\beta - I) QB\tilde{A}^T (Q^\beta + (\beta - I)C)\tilde{A}A \\
&\quad - \tilde{A}^T (Q^\beta + (\beta - I)C)\tilde{A}BQ + QBQ \Bigg) x \\
&= x^T \Bigg(\tilde{A}^T R(Q) - (\beta - I) QB\tilde{A}^T R(Q) \\
&\quad + QB\tilde{A}^T (Q^\beta + (\beta - I)C) - QB\tilde{A}^T Q + (\beta - I) QB\tilde{A}^T A^T Q \\
&\quad + QB\tilde{A}^T (Q^\beta + (\beta - I)C)\tilde{A} - QB\tilde{A}^T (Q^\beta + (\beta - I)C) \\
&\quad - \tilde{A}^T (Q^\beta + (\beta - I)C)\tilde{A}BQ + QBQ \Bigg) x \\
&= x^T \Bigg(\tilde{A}^T R(Q) - (\beta - I) QB\tilde{A}^T R(Q) - QB\tilde{A}^T Q + QB\tilde{A}^T Q \\
&\quad - QBQ + QB\tilde{A}^T (Q^\beta + (\beta - I)C)\tilde{A} - \tilde{A}^T (Q^\beta + (\beta - I)C)\tilde{A}BQ + QBQ \Bigg) x \\
&= x^T \left(\tilde{A}^T R(Q) - (\beta - I) QB\tilde{A}^T R(Q) \right) x, \quad on \quad [a,b],
\end{aligned}
$$

and

$$
\begin{aligned}
& u^T \left((\beta - I) B\tilde{A}^T D_\beta Q\tilde{A} + (\beta - I) B\tilde{A}^T Q\tilde{A}A + B\tilde{A}^T Q + (\beta - I) B\tilde{A}^T C\tilde{A} - DQ \right) x \\
&= u^T \Bigg((\beta - I) B\tilde{A}^T R(Q) - (\beta - I) B\tilde{A}^T D_\beta Q - (\beta - I) B\tilde{A}^T C - (\beta - I) B\tilde{A}^T A^T Q \\
&\quad - (\beta - I) B\tilde{A}^T (Q^\beta + (\beta - I)C)\tilde{A}(A + BQ) + (\beta - I) B\tilde{A}^T D_\beta Q\tilde{A} \\
&\quad + (\beta - I) B\tilde{A}^T Q\tilde{A}A + B\tilde{A}^T Q + (\beta - I) B\tilde{A}^T C\tilde{A} - DQ \Bigg) x
\end{aligned}
$$

$$
\begin{aligned}
&= u^T \Bigg((\beta - I)B\tilde{A}^T R(Q) + (\beta - I)^2 B\tilde{A}^T D_\beta Q\tilde{A}A + (\beta - I)^2 B\tilde{A}^T C\tilde{A}A \\
&\quad -(\beta - I)B\tilde{A}^T A^T Q - (\beta - I)B\tilde{A}^T (Q^\beta + (\beta - I)C)\tilde{A}(A + BQ) \\
&\quad +(\beta - I)B\tilde{A}^T Q\tilde{A}A + B\tilde{A}^T Q \\
&\quad -BQ + (\beta - I)B\tilde{A}^T (Q^\beta + (\beta - I)C)\tilde{A}BQ \Bigg) x \\
&= u^T \Bigg((\beta - I)B\tilde{A}^T R(Q) + (\beta - I)B\tilde{A}^T Q^\beta \tilde{A}A - (\beta - I)B\tilde{A}^T Q\tilde{A}A \\
&\quad +(\beta - I)^2 B\tilde{A}^T C\tilde{A}A - B(\tilde{A}^T - I)Q \\
&\quad -(\beta - I)B\tilde{A}^T (Q^\beta + (\beta - I)C)\tilde{A}(A + BQ) \\
&\quad +(\beta - I)B\tilde{A}^T Q\tilde{A}A + B\tilde{A}^T Q \\
&\quad -BQ + (\beta - I)B\tilde{A}^T (Q^\beta + (\beta - I)C)\tilde{A}BQ \Bigg) x \\
&= u^T \Bigg((\beta - I)B\tilde{A}^T R(Q) + (\beta - I)B\tilde{A}^T (Q^\beta + (\beta - I)C)\tilde{A}A \\
&\quad -(\beta - I)B\tilde{A}^T (Q^\beta + (\beta - I)C)\tilde{A}(A + BQ) \\
&\quad +(\beta - I)B\tilde{A}^T (Q^\beta + (\beta - I)C)\tilde{A}BQ \Bigg) x \\
&= u^T \left((\beta - I)B\tilde{A}^T R(Q) \right) x \quad on \quad [a, b],
\end{aligned}
$$

and

$$
\begin{aligned}
&\left(x^T \left((\beta - I)\tilde{A}^T D_\beta Q\tilde{A}B + \tilde{A}^T Q\tilde{A}B + (\beta - I)\tilde{A}^T C\tilde{A}B - QD \right) u \right)^T \\
&= u^T \left((\beta - I)B\tilde{A}^T D_\beta Q\tilde{A} + (\beta - I)B\tilde{A}^T Q\tilde{A}A + B\tilde{A}^T Q + (\beta - I)B\tilde{A}^T C\tilde{A} - DQ \right) x \\
&= u^T \left((\beta - I)B\tilde{A}^T R(Q) \right) x \quad on \quad [a, b],
\end{aligned}
$$

and

$$u^T\left((\beta-I)^2 B\tilde{A}^T D_\beta Q\tilde{A}B+(\beta-I)B\tilde{A}^T Q\tilde{A}B+(\beta-I)^2 B\tilde{A}C\tilde{A}B-B+D\right)u$$

$$=u^T\Big((\beta-I)B\tilde{A}^T(Q^\beta-Q)\tilde{A}B+(\beta-I)B\tilde{A}^T Q\tilde{A}B+(\beta-I)^2 B\tilde{A}^T C\tilde{A}B-B+D\Big)u$$

$$=u^T\Big((\beta-I)B\tilde{A}^T Q^\beta\tilde{A}B-(\beta-I)B\tilde{A}^T Q\tilde{A}B+(\beta-I)B\tilde{A}^T Q\tilde{A}B+(\beta-I)^2 B\tilde{A}^T C\tilde{A}B-B+D\Big)$$

$$=u^T\Big((\beta-I)B\tilde{A}^T Q^\beta\tilde{A}B+(\beta-I)^2 B\tilde{A}^T C\tilde{A}B-B+B-(\beta-I)B\tilde{A}^T(Q^\beta+(\beta-I)C)\tilde{A}B\Big)u$$

$$=0\quad on\quad [a,b].$$

Consequently,

$$D_\beta(x^T Qx)+(x^\beta)^T Cx^\beta-u^T Bu+z^T Dz$$

$$=x^T\left(\tilde{A}^T R(Q)-(\beta-I)QB\tilde{A}^T R(Q)\right)x$$

$$+2u^T\left((\beta-I)B\tilde{A}^T R(Q)\right)x\quad on\quad [a,b].$$

Now we will prove (4.14). We have

$$x+(\beta-I)Dz-(\beta-I)^2 B\tilde{A}^T R(Q)x$$

$$=x+(\beta-I)\Big(B-(\beta-I)B\tilde{A}^T(Q^\beta+(\beta-I)C)\tilde{A}B\Big)(u-Qx)$$

$$-(\beta-I)^2 B\tilde{A}^T\Big(D_\beta Q+C+A^T Q+(Q^\beta+(\beta-I)C)\tilde{A}(A+BQ)\Big)x$$

$$=x+(\beta-I)Bu-(\beta-I)BQx-(\beta-I)^2 B\tilde{A}^T(Q^\beta+(\beta-I)C)\tilde{A}Bu$$

$$+(\beta-I)^2 B\tilde{A}^T(Q^\beta+(\beta-I)C)\tilde{A}BQx$$

$$-(\beta-I)B\tilde{A}^T(Q^\beta-Q+(\beta-I)C)x-(\beta-I)^2 B\tilde{A}^T A^T Qx$$

$$-(\beta-I)^2 B\tilde{A}^T(Q^\beta+(\beta-I)C)\tilde{A}Ax$$

$$-(\beta-I)^2 B\tilde{A}^T(Q^\beta+(\beta-I)C)\tilde{A}BQx$$

$$= -(\beta - I)^2 B\tilde{A}^T(Q^\beta + (\beta - I)C)\tilde{A}(Ax + Bu)$$

$$+x + (\beta - I)Bu - (\beta - I)BQx - (\beta - I)^2 B\tilde{A}^T A^T Qx$$

$$-(\beta - I)B\tilde{A}^T(Q^\beta - Q + (\beta - I)C)x$$

$$= -(\beta - I)^2 B\tilde{A}^T(Q^\beta + (\beta - I)C)D_\beta x - (\beta - I)B\tilde{A}^T(Q^\beta + (\beta - I)C)x$$

$$+(\beta - I)B\tilde{A}^T Qx + x + (\beta - I)Bu - (\beta - I)BQx - (\beta - I)^2 B\tilde{A}^T A^T Qx$$

$$= -(\beta - I)B\tilde{A}^T(Q^\beta + (\beta - I)C)x^\beta$$

$$+(\beta - I)B\tilde{A}^T(I - (\beta - I)A^T)Qx - (\beta - I)BQx + x + (\beta - I)Bu$$

$$= -(\beta - I)B\tilde{A}^T(Q^\beta + (\beta - I)C)x^\beta + (\beta - I)BQx - (\beta - I)BQx + x + (\beta - I)Bu$$

$$= -(\beta - I)B\tilde{A}^t(Q^\beta + (\beta - I)C)x^\beta + x + (\beta - I)Bu$$

$$= -(\beta - I)B\tilde{A}^T(Q^\beta + (\beta - I)C)x^\beta + \tilde{A}^{-1}(\tilde{A}x + (\beta - I)\tilde{A}Bu)$$

$$= -(\beta - I)B\tilde{A}^T(Q^\beta + (\beta - I)C)x^\beta + \tilde{A}^{-1}x^\beta$$

$$= \left(\tilde{A}^{-1} - (\beta - I)B\tilde{A}^T(Q^\beta + (\beta - I)C)\right)x^\beta \quad on \quad [a,b].$$

This completes the proof. □

Lemma 4.2 *For any two matrices V and W, we have*

$$KerV \subseteq KerW \quad iff \quad W = WV^\dagger V \quad iff \quad W^\dagger = V^\dagger V W^\dagger.$$

Proof Let $W = WV^\dagger V$. Let also, $x \in KerV$ be arbitrarily chosen. Then

$$Vx = 0.$$

Hence,

$$Wx = WV^\dagger Vx$$

$$= 0.$$

Therefore, $x \in KerW$. Because $x \in KerV$ was arbitrarily chosen and for it we get that it is an element of $KerW$, we conclude that

$$KerV \subseteq KerW.$$

Now we suppose that

$$KerV \subseteq KerW.$$

Let x be such that $W^T x$ is defined and there exists

$$x = Vd_1 + d_2$$

with $d_2 \in KerV^T$. Therefore,

$$\begin{aligned} W^T x &= V^T (Vd_1 + d_2) \\ &= V^T V d_1 + V^T d_2 \\ &= V^T V d_1 \\ &= V^T V V^\dagger V d_1 \\ &= V^T \left(V V^\dagger\right)^T V d_1 \\ &= V^T (V^\dagger)^T V^T V d_1 \\ &= V^T (V^\dagger)^T W^T x \\ &= (W V^\dagger V)^T x. \end{aligned}$$

Therefore,

$$W = W V^\dagger V.$$

Note that

$$\begin{aligned} W^\dagger &= (W V^\dagger V)^\dagger \\ &= V^\dagger (V^\dagger)^\dagger W^\dagger \\ &= V^\dagger V W^\dagger. \end{aligned}$$

This completes the proof. □

Lemma 4.3 *Suppose that*

$$D_\beta X = AX^\beta + BU \tag{4.15}$$

and $KerX^\beta \subseteq KerX$ *hold at* $t \in [a,b]$. *Then*

$$X = X(X^\beta)^\dagger X^\beta, \quad X^\dagger = (X^\beta)^\dagger X^\beta X^\dagger \tag{4.16}$$

and

$$(\beta - I)X^\beta(X^\beta)^\dagger \tilde{A}B = (\beta - I)\tilde{A}B.$$

Proof We apply Lemma 4.2 for $W = X$ and $V = X^\beta$ and we get (4.16). By (4.15), we get

$$X^\beta = \tilde{A}X + (\beta - I)\tilde{A}BU$$

and

$$Ker(X^\beta)^T \subseteq Ker((\beta - I)B\tilde{A}^T) \quad at \quad t.$$

From this and, we obtain

$$(\beta - I)B\tilde{A}^T = (\beta - I)B\tilde{A}^T(X^\beta)^{\dagger T}(X^\beta)^T,$$

whereupon

$$(\beta - I)X^\beta(X^\beta)^\dagger \tilde{A}B = (\beta - I)\tilde{A}B.$$

This completes the proof. □

Lemma 4.4 *Let* $t \in [a,b]$. *Suppose that* $X, U : [a,b] \to \mathbb{R}^{n\times n}$ *are matrices such that* $KerX^\beta \subseteq KerX$ *and*

$$D_\beta X = AX^\beta + BU$$

hold at t. Let also, U and Q be differentiable at t with

$$QX = UX^\dagger X$$

symmetric on $[t, \beta(t)]$.

1. *If t is right-scattered, or if t is right-dense with X nonsingular at t, then*

$$R(Q)X = (D_\beta U + CX^\beta + A^T U)X^\dagger X.$$

2. *If*

$$D_\beta U = -CX^\beta - A^T U \quad at \quad t,$$

and

$$X^T U = U^T X \quad at \quad t,$$

then

$$D = X(X^\beta)^T \tilde{A} = B - (\beta - I)B\tilde{A}^T(Q^\beta + (\beta - I)C)\tilde{A}B$$

is symmetric at t and

$$X^T \tilde{A}^T R(Q)X = (\beta - I)B\tilde{A}^T R(Q)X = 0$$

at t. Moreover, if t is right-scattered, or if t is right-dense with X nonsingular at t, then we have

$$R(Q)X = 0 \quad at \quad t,$$

or

$$R(Q) = 0 \quad at \quad t,$$

respectively.

Proof 1. Let

$$Z = U - QX.$$

Then

$$QX = U - Z$$

and

$$\begin{aligned} ZX^{\dagger}X &= (U - QX)X^{\dagger}X \\ &= (U - UX^{\dagger}X)X^{\dagger}X \\ &= UX^{\dagger}X - UX^{\dagger}XX^{\dagger}X \\ &= UX^{\dagger}X - UX^{\dagger}X \\ &= 0 \quad at \quad t. \end{aligned}$$

Then

$$\begin{aligned} R(Q)X &= \left(D_{\beta}Q + C + A^{T}Q + (Q^{\beta} + (\beta - I)C)\tilde{A}(A + BQ)\right)X \\ &= \left(D_{\beta}Q + C + A^{T}Q + (Q^{\beta} + (\beta - I)C)\tilde{A}(A + BQ)\right)XX^{\dagger}X \\ &= D_{\beta}QXX^{\dagger}X + CXX^{\dagger}X + A^{T}QXX^{\dagger}X \\ &\quad + (Q^{\beta} + (\beta - I)C)\tilde{A}(A + BQ)XX^{\dagger}X \\ &= D_{\beta}(QX)X^{\dagger}X - Q^{\beta}D_{\beta}XX^{\dagger}X + CXX^{\dagger}X \\ &\quad + A^{T}(U - Z)X^{\dagger}X \\ &\quad + (Q^{\beta} + (\beta - I)C)\tilde{A}(AX + BU - BZ)X^{\dagger}X \end{aligned}$$

$$= D_\beta(U-Z)X^\dagger X - Q^\beta D_\beta XX^\dagger X + CXX^\dagger X$$

$$+A^T(U-Z)X^\dagger X$$

$$+(Q^\beta + (\beta - I)C)\tilde{A}(AX + BU - BZ)X^\dagger X$$

$$= D_\beta UX^\dagger X - Z^{D_\beta}X^\dagger X - Q^\beta D_\beta XX^\dagger X + CXX^\dagger X$$

$$+A^T UX^\dagger X + (Q^\beta + (\beta - I)C)\tilde{A}(AX + BU)X^\dagger X$$

$$= D_\beta UX^\dagger X - D_\beta ZX^\dagger X - Q^\beta D_\beta XX^\dagger X + CXX^\dagger X$$

$$+A^T UX^\dagger X + (Q^\beta + (\beta - I)C)D_\beta XX^\dagger X$$

$$= D_\beta UX^\dagger X - D_\beta ZX^\dagger X - Q^\beta D_\beta XX^\dagger X + CXX^\dagger X$$

$$+A^T UX^\dagger X + Q^\beta D_\beta XX^\dagger X + C(X^\beta - X)X^\dagger X$$

$$= D_\beta UX^\dagger X - D_\beta ZX^\dagger X + A^T UX^\dagger X + CX^\beta X^\dagger X$$

$$= \left(D_\beta U + CX^\beta + A^T U\right)X^\dagger X - D_\beta ZX^\dagger X.$$

If $\beta(t) - t > 0$, then

$$\begin{aligned} D_\beta ZX^\dagger X &= \frac{1}{(\beta - I)}(Z^\beta - Z)X^\dagger X \\ &= \frac{1}{(\beta - I)}Z^\beta X^\dagger X \\ &= \frac{1}{(\beta - I)}Z^\beta (X^\beta)^\dagger X^\beta X^\dagger X \\ &= \frac{1}{(\beta - I)}(ZX^\dagger X)^\beta X^\dagger X \\ &= 0. \end{aligned}$$

If $\beta(t) = t$, then

$$X^\dagger = X^{-1}$$

and

$$Z = ZX^{-1}X$$

$$= ZX^{\dagger}X$$

$$= 0.$$

Therefore,

$$R(Q)X = \left(D_{\beta}U + CX^{\beta} + A^{T}U\right)X^{\dagger}X.$$

2. Note that

$$X^{T}QX = X^{T}UX^{\dagger}X$$

$$= U^{T}XX^{\dagger}X$$

$$= U^{T}X.$$

Then

$$D = X(X^{\beta})^{\dagger}\tilde{A}B$$

$$= \left(\left(-(\beta - I)^{2}B\tilde{A}^{T}C + \tilde{A}^{-1}\right)X^{\beta} - (\beta - I)B\tilde{A}^{T}U^{\beta}\right)(X^{\beta})^{\dagger}\tilde{A}B$$

$$= -(\beta - I)^{2}B\tilde{A}^{T}C\tilde{A}B + B - (\beta - I)B\tilde{A}^{T}U^{\beta}(X^{\beta})^{\dagger}\tilde{A}B$$

$$= -(\beta - I)^{2}B\tilde{A}^{T}C\tilde{A}B + B$$

$$-(\beta - I)B\tilde{A}^{T}(X^{\beta})^{\dagger T}(X^{\beta})^{T}Q^{\beta}X^{\beta}(X^{\beta})^{\dagger}\tilde{A}B$$

$$= B - (\beta - I)B\tilde{A}^{T}(Q^{\beta} + (\beta - I)C)\tilde{A}B,$$

i.e., D is symmetric. Note that

$$X^{T}Z = 0$$

and

$$\begin{aligned}\tilde{A}X &= X^{\beta} - (\beta - I)\tilde{A}BU \\ &= X^{\beta}(X^{\beta})^{\dagger}X^{\beta} - (\beta - I)X^{\beta}(X^{\beta})^{\dagger}\tilde{A}BU \\ &= X^{\beta}(X^{\beta})^{\dagger}(X^{\beta} - (\beta - I)\tilde{A}BU) \\ &= X^{\beta}(X^{\beta})^{\dagger}\tilde{A}X.\end{aligned}$$

Now we apply the computations of 1 and we obtain

$$\begin{aligned}X^T\tilde{A}^T R(Q)X &= -X^T\tilde{A}^T D_{\beta}ZX^{\dagger}X \\ &= -X^T\tilde{A}^T(X^{\beta})^{\dagger T}(X^{\beta})^T D_{\beta}ZX^{\dagger}X \\ &= -X^T\tilde{A}^T(X^{\beta})^{\dagger T}D_{\beta}(X^TZ)X^{\dagger}X \\ &\quad +X^T\tilde{A}^T(X^{\beta})^{\dagger T}(D_{\beta}X)^T ZX^{\dagger}X \\ &= 0\end{aligned}$$

and

$$\begin{aligned}(\beta - I)B\tilde{A}^T R(Q)X &= -(\beta - I)B\tilde{A}^T(X^{\beta})^{\dagger T}D_{\beta}(X^TZ)X^{\dagger}X \\ &\quad +(\beta - I)B\tilde{A}^T(X^{\beta})^{\dagger T}(D_{\beta}X)^T ZX^{\dagger}X \\ &= 0.\end{aligned}$$

If $\beta(t) > t$, then, using 1, we get

$$R(Q)X = 0 \quad at \quad t.$$

If $t = \beta(t)$ is right-dense with X nonsingular at T, using 1, we obtain

$$R(Q) = 0 \quad at \quad t.$$

This completes the proof. □

Below we suppose:

(A) Let (X,U) and $(\tilde{X},\tilde{U})$ be normalized conjoined basis of (4.6) such that

$$KerX^{\beta} \subseteq KerX$$

holds on $[a,b]$ and X is nonsingular at all dense points of $(a,b]$.

We define the matrix

$$Q = UX^{\dagger} + (UX^{\dagger}\tilde{X} - \tilde{U})(I - X^{\dagger}X)U^{T} \quad on \quad [a,b]. \tag{4.17}$$

Because (X,U) and $(\tilde{X},\tilde{U})$ is a normalized conjoined basis, we have

$$X^{T}U - U^{T}X = 0, \quad \tilde{X}^{T}\tilde{U} - \tilde{U}^{T}\tilde{X} = 0 \quad on \quad [a,b],$$

and

$$X^{T}\tilde{U} - U^{T}\tilde{X} = I \quad on \quad [a,b].$$

Note that

$$QX = UX^{\dagger}X + (UX^{\dagger}\tilde{X} - \tilde{U})(I - X^{\dagger}X)U^{T}X \quad on \quad [a,b].$$

Since

$$\begin{aligned}
&\left((UX^{\dagger}\tilde{X} - \tilde{U})(I - X^{\dagger}X)U^{T}X\right)^{T} \\
&= X^{T}U(I - X^{\dagger}X)(UX^{\dagger}\tilde{X} - \tilde{U})^{T} \\
&= U^{T}X(I - X^{\dagger}X)(UX^{\dagger}\tilde{X} - \tilde{U})^{T} \\
&= U^{T}(X - XX^{\dagger}X)(UX^{\dagger}\tilde{X} - \tilde{U})^{T} \\
&= O \quad on \quad [a,b].
\end{aligned}$$

Therefore,

$$QX = UX^{\dagger}X \quad on \quad [a,b].$$

Next,

$$\begin{aligned}
Q &= UX^{\dagger} + (UX^{\dagger}\tilde{X} - \tilde{U})(U^{T} - X^{\dagger}XU^{T}) \\
&= UX^{\dagger} + UX^{\dagger}\tilde{X}U^{T} - \tilde{U}U^{T} - UX^{\dagger}\tilde{X}X^{\dagger}XU^{T} \\
&\quad + \tilde{U}X^{\dagger}XU^{T}
\end{aligned}$$

$$
\begin{aligned}
&= -\tilde{U}U^T - UX^\dagger \tilde{X} X U^T + UX^\dagger \\
&\quad + UX^\dagger \tilde{X} U^T + \tilde{U} X^\dagger X U^T \\
&= -\tilde{U}U^T - UX^\dagger \tilde{X} X^\dagger X U^T + \tilde{U} X^\dagger X U^T \\
&\quad + UX^\dagger (I + \tilde{X} U^T) \\
&= -\tilde{U}U^T - UX^\dagger \tilde{X} X^\dagger X U^T + \tilde{U} X^\dagger U^T \\
&\quad + UX^\dagger X \tilde{U}^T \quad on \quad [a,b].
\end{aligned}
$$

By Theorem 4.8, item 1, we have that $U\tilde{U}^T$ is symmetric. Then

$$
\begin{aligned}
(U\tilde{U}^T)^T &= U\tilde{U}^T \\
&= \tilde{U}U^T,
\end{aligned}
$$

hence,

$$
\begin{aligned}
\left(\tilde{U}U^T\right)^T &= \left(U\tilde{U}^T\right)^T \\
&= U\tilde{U}^T \\
&= \tilde{U}U^T.
\end{aligned}
$$

Next, again by Theorem 4.8, 1, we have that $X\tilde{X}^T$ is symmetric and

$$
\begin{aligned}
\left(X\tilde{X}^T\right)^T &= X\tilde{X}^T \\
&= \tilde{X}X^T,
\end{aligned}
$$

and

$$
\begin{aligned}
\left(UX^\dagger \tilde{X} X^\dagger X U^T\right)^T &= UX^\dagger X \tilde{X}^T \left(X^\dagger\right)^T U^T \\
&= UX^\dagger \tilde{X} X^T (X^\dagger)^T U^T \\
&= UX^\dagger \tilde{X} (X^\dagger X)^T U^T \\
&= UX^\dagger \tilde{X} X^\dagger X U^T \quad on \quad [a,b].
\end{aligned}
$$

Therefore,

$$\begin{aligned} Q^T &= -(\tilde{U}U^T)^T - \left(UX^{\dagger}\tilde{X}X^{\dagger}XU^T\right)^T \\ &\quad + \left(\tilde{U}X^{\dagger}XU^T\right)^T + \left(UX^{\dagger}X\tilde{U}^T\right)^T \\ &= -\tilde{U}U^T - UX^{\dagger}\tilde{X}X^{\dagger}XU^T \\ &\quad + UX^{\dagger}X\tilde{U}^T + \tilde{U}X^{\dagger}XU^T \\ &= Q \quad on \quad [a,b], \end{aligned}$$

i.e., Q is symmetric on $[a,b]$. Note that Q is differentiable on $[a,b]^{\kappa}$ and if t is right-dense, then X is nonsingular at t and $Q = UX^{-1}$ at t. Also, if (X,U) has no focal point in $(a,b]$, then (A) holds. Now we consider the functional

$$\mathscr{F}_0(x,u) = \int_a^b \left(u^T Bu - (x^{\beta})^T Cx^{\beta}\right)(t)d_{\beta}t.$$

Theorem 4.12 *Suppose that the hypothesis* (A) *holds. Let an admissible* (x,u) *with* $x(a) \in ImX(a)$ *be given. Then* $x \in ImX$ *on* $[a,b]$ *and*

$$\begin{aligned} \mathscr{F}_0(x,u) &= (x(b))^T Q(b)x(b) - (x(a))^T Q(a)x(a) \\ &\quad + \int_a^b \left(z^T Dz\right)(t)d_{\beta}t, \end{aligned}$$

and

$$x + (\beta - I)Dz = \left(\tilde{A}^{-1} - (\beta - I)B\tilde{A}^T\left(Q^{\beta} + (\beta - I)C\right)\right)x^{\beta} \quad on \quad [a,b],$$

where Q *is given by* (4.17),

$$z = u - Qx = u - UX^{\dagger}x \quad on \quad [a,b],$$

and

$$D = B - (\beta - I)B\tilde{A}^T\left(Q^{\beta} + (\beta - I)C\right)\tilde{A}B \quad on \quad [a,b].$$

Proof Let (X,U) and $(\tilde{X},\tilde{U})$ be normalized conjoined basis of (4.6) satisfying (A). We have that Q is differentiable on $[a,b]$. We will prove that $x(t) \in ImX(t)$ for all $t \in [a,b]$. Define the statement

$$E(t) = \{x(t) \in ImX(t), \quad t \in [a,b]\}.$$

We have

1. $E(a)$ holds.
2. Let $\beta(t) > t$ and $E(t)$ hold. Then by Theorem 4.10, it follows that $E^{\beta}(t)$ holds.
3. Let $\beta(t) = tt$ and $E(t)$ hold. Then X is nonsingular at t and there exists a neighborhood V of t such that X is invertible at any $s \in V$, $s > t$. Therefore,
$$d(s) = (X(s))^{-1}x(s) \in \mathbb{R}^n$$
and
$$x(s) = X(s)d(s) \in ImX(s).$$
Therefore, $E(s)$ holds for all $s \in V$, $s > t$.
4. Let t be left-dense and $E(s)$ is true for all $s \in [a,t)$. Then X is nonsingular at t. Hence,
$$d(t) = (X(t))^{-1}x(t) \in \mathbb{R}^n$$
and
$$x(t) = X(t)d(t) \in ImX(t).$$

Consequently $x \in ImX$ on $[a,b]$. By Lemma 4.4, item 2, it follows that

$$\begin{aligned} D &= X(X^{\beta})^{\dagger}\tilde{A}B \\ &= B - (\beta - I)B\tilde{A}^T(Q^{\beta} + (\beta - I)C)\tilde{A}B \end{aligned}$$

is symmetric at t and

$$\begin{aligned} X^T\tilde{A}^T R(Q)X &= (\beta - I)B\tilde{A}^T R(Q)X \\ &= 0 \quad on \quad [a,b]. \end{aligned}$$

By Lemma 4.1, it follows that

$$x + (\beta - I)Dz = \left(\tilde{A}^{-1} - (\beta - I)B\tilde{A}^T(Q^{\beta} + (\beta - I)C)\right)x^{\beta} \quad on \quad [a,b],$$

and

$$D_{\beta}\left(x^T Qx\right) + (x^{\beta})^T Cx^{\beta} - u^T Bu + z^T Dz = 0 \quad on \quad [a,b].$$

Hence,

$$\begin{aligned} \mathscr{F}_0(x,u) &= \int_a^b \left(u^T Bu - (x^{\beta})^T Cx^{\beta}\right)(t)D_{\beta}t \\ &= \int_a^b D_{\beta}\left(x^T Qx\right)(t)D_{\beta}t + \int_a^b (z^T Dz)(t)D_{\beta}t \\ &= (x^T Qx)(b) - (x^T Qx)(a) + \int_a^b (z^T Dz)(t)D_{\beta}t. \end{aligned}$$

Let now $t \in [a,b]$ and

$$x(t) = X(t)d(t)$$

for some $d(t) \in \mathbb{R}^n$. Then

$$\begin{aligned} Qx(t) &= QXd(t) \\ &= UX^{\dagger}Xd(t) \\ &= UX^{\dagger}x(t). \end{aligned}$$

Therefore,

$$\begin{aligned} z &= u - Qx \\ &= u - UX^{\dagger}x \quad on \quad [a,b]. \end{aligned}$$

This completes the proof. □

4.6 "Big" Linear Hamiltonian Systems

We define $2n \times 2n$ matrices A^*, B^*, C^* as follows

$$A^* = \begin{pmatrix} O & O \\ O & A \end{pmatrix}, \quad B^* \begin{pmatrix} O & O \\ O & B \end{pmatrix}, \quad C^* = \begin{pmatrix} O & O \\ O & C \end{pmatrix} \quad on \quad [a,b].$$

Then

$$\begin{aligned} I - (\beta - I)A^* &= \begin{pmatrix} I & O \\ O & I \end{pmatrix} - (\beta - I) \begin{pmatrix} O & O \\ O & A \end{pmatrix} \\ &= \begin{pmatrix} I & O \\ O & I - (\beta - I)A \end{pmatrix} \quad on \quad [a,b]. \end{aligned}$$

Hence, $I - (\beta - I)A^*$ is invertible on $[a,b]$ with

$$\begin{aligned} \tilde{A}^* &= (I - (\beta - I)A^*)^{-1} \\ &= \begin{pmatrix} I & O \\ O & \tilde{A} \end{pmatrix} \quad on \quad [a,b]. \end{aligned}$$

Also, $A^*, B^*, C^* \in \mathscr{C}_\beta([a,b])$, B^* and C^* are symmetric. For $n \times n$ matrices $\tilde{X}$ and $\tilde{U}$ we define the matrices

$$X^* = \begin{pmatrix} O & I \\ X & \tilde{X} \end{pmatrix}, \quad U^* = \begin{pmatrix} I & O \\ U & \tilde{U} \end{pmatrix} \quad on \quad [a,b].$$

Consider the "big" linear Hamiltonian system

$$\begin{cases} D_\beta(X^*) = A^*(X^\star)^\beta + B^* U^* \\ \\ D_\beta(U^*) = -C^*(X^*)^\beta - (A^*)^T U^* \quad on \quad [a,b]. \end{cases} \tag{4.18}$$

Suppose that the hypothesis (A) holds.

Theorem 4.13 (X,U) *and* $(\tilde{X},\tilde{U})$ *are normalized conjoined bases of the system* (4.6) *if and only if* (X^*,U^*) *is a conjoined basis of the system* (4.18).

Proof 1. Let (X,U) and $(\tilde{X},\tilde{U})$ be normalized conjoined bases of the system (4.6). Then

$$\begin{cases} X^T U - U^T X = O \\ \\ \tilde{X}^T \tilde{U} - \tilde{U}^T \tilde{X} = O \\ \\ X^T \tilde{U} - U^T \tilde{X} = I \quad on \quad [a,b]. \end{cases} \tag{4.19}$$

Then

$$\begin{aligned} (X^*)^T U^* - (U^*)^T X^* &= \begin{pmatrix} O & X^T \\ I & \tilde{X}^T \end{pmatrix} \begin{pmatrix} I & O \\ U & \tilde{U} \end{pmatrix} \\ &\quad - \begin{pmatrix} I & U^T \\ O & \tilde{U}^T \end{pmatrix} \begin{pmatrix} O & I \\ X & \tilde{X} \end{pmatrix} \\ &= \begin{pmatrix} X^T U & X^T \tilde{U} \\ I + \tilde{X}^T U & \tilde{X}^T \tilde{U} \end{pmatrix} \\ &\quad - \begin{pmatrix} U^T X & I - U^T \tilde{X} \\ \tilde{U}^T X & \tilde{U}^T \tilde{X} \end{pmatrix} \\ &= \begin{pmatrix} X^T U - U^T X & X^T \tilde{U} - I + U^T \tilde{X} \\ I + \tilde{X}^T U - \tilde{U}^T X & \tilde{X}^T \tilde{U} - \tilde{U}^T \tilde{X} \end{pmatrix} \\ &= \begin{pmatrix} O & O \\ O & O \end{pmatrix} \\ &= O. \end{aligned}$$

Therefore, (X^*,U^*) is a conjoined basis of (4.18).

2. Let (X^*, U^*) be a conjoined basis of (4.18). Then, using the computations in the previous step, we get (4.19). This completes the proof. □

Let

$$Q = UX^\dagger + \left(UX^\dagger \tilde{X} - \tilde{U}\right)\left(I - X^\dagger X\right)U^T,$$

$$\tilde{Q} = X^\dagger + X^\dagger \tilde{X}\left(I - X^\dagger X\right)U^T,$$

$$Q^* = \begin{pmatrix} -X^\dagger \tilde{X} X^\dagger X & \tilde{Q} \\ \tilde{Q}^T & Q \end{pmatrix}.$$

By the previous section, we have that Q is symmetric with

$$QX = UX^\dagger X \quad on \quad [a,b].$$

Also, we have

$$\begin{aligned} \left(X^\dagger \tilde{X} X^\dagger X\right)^T &= \left(X^\dagger X\right)^T \left(X^\dagger \tilde{X}\right)^T \\ &= X^\dagger X \left(\tilde{X}\right)^T \left(X^\dagger\right)^T \\ &= X^\dagger \tilde{X} X^T \left(X^\dagger\right)^T \\ &= X^\dagger \tilde{X} \left(X^\dagger X\right)^T \\ &= X^\dagger \tilde{X} X^\dagger X. \end{aligned}$$

Therefore,

$$\begin{aligned} (Q^*)^T &= \begin{pmatrix} -\left(X^\dagger \tilde{X} X^\dagger X\right)^T & \tilde{Q} \\ \tilde{Q}^T & Q \end{pmatrix} \\ &= \begin{pmatrix} -X^\dagger \tilde{X} X^\dagger X & \tilde{Q} \\ \tilde{Q}^T & Q \end{pmatrix} \\ &= Q^* \quad on \quad [a,b], \end{aligned}$$

i.e., Q^* is symmetric on $[a,b]$. Next,

$$\begin{aligned} Q^* X^* &= \begin{pmatrix} -X^\dagger \tilde{X} X^\dagger X & \tilde{Q} \\ \tilde{Q}^T & Q \end{pmatrix} \begin{pmatrix} O & I \\ X & \tilde{X} \end{pmatrix} \\ &= \begin{pmatrix} \tilde{Q}X & -X^\dagger \tilde{X} X^\dagger X + \tilde{Q}\tilde{X} \\ QX & \tilde{Q}^T + Q\tilde{X} \end{pmatrix}. \end{aligned}$$

Consider

$$\tilde{Q}X = X^\dagger X + X^\dagger \tilde{X}(I - X^\dagger X)U^T X.$$

Since

$$\begin{aligned}
\left(X^\dagger \tilde{X}(I - X^\dagger X)U^T X\right)^T &= X^T U(I - X^\dagger X)(X^\dagger \tilde{X})^T \\
&= U^T X(I - X^\dagger X)(X^\dagger \tilde{X})^T \\
&= U^T (X - XX^\dagger X)(X^\dagger \tilde{X})^T \\
&= U^T (X - X)(X^\dagger \tilde{X})^T \\
&= 0,
\end{aligned}$$

we get

$$\tilde{Q}X = X^\dagger X.$$

Next,

$$\begin{aligned}
-X^\dagger \tilde{X} X^{+} X + \tilde{Q}\tilde{X} &= -X^\dagger \tilde{X} X^\dagger X + \left(X^\dagger + X^\dagger \tilde{X}(I - X^\dagger X)U^T\right)\tilde{X} \\
&= -X^\dagger \tilde{X} X^\dagger X + \left(X^\dagger \tilde{X} + X^\dagger \tilde{X}(I - X^\dagger \tilde{X})(I - X^\dagger X)(X^T \tilde{U} - I)\right) \\
&= -X^\dagger \tilde{X} X^\dagger X + X^\dagger \tilde{X} - X^\dagger \tilde{X} + X^\dagger \tilde{X} X^\dagger X \\
&\quad + X^\dagger \tilde{X}(I - X^\dagger X)X^T \tilde{U} \\
&= X^\dagger \tilde{X}(I - X^\dagger X)X^T \tilde{U},
\end{aligned}$$

and using that

$$\begin{aligned}
\left((I - X^\dagger X)X^T\right)^T &= X(I - X^\dagger X) \\
&= X - XX^\dagger X \\
&= X - X \\
&= 0,
\end{aligned}$$

we get

$$-X^{\dagger}\tilde{X}X^{\dagger}X+\tilde{Q}\tilde{X}=O.$$

Also,

$$\begin{aligned}
\tilde{Q}^T &= \left(X^{\dagger}\right)^T + U\left(X^{\dagger}\tilde{X}\right)^T - UX^{\dagger}\tilde{X}X^{\dagger}X\\
&= (X^{\dagger})^T + U\tilde{X}^T(X^{\dagger})^T - UX^{\dagger}\tilde{X}X^{\dagger}X\\
&= (X^{\dagger})^T + (\tilde{U}X^T - I)(X^{\dagger})^T - UX^{\dagger}\tilde{X}X^{\dagger}X\\
&= \tilde{U}(X^{\dagger}X)^T - UX^{\dagger}\tilde{X}X^{\dagger}X\\
&= \tilde{U}X^{\dagger}X - UX^{\dagger}\tilde{X}X^{\dagger}X,\\
Q\tilde{X} &= UX^{\dagger}\tilde{X} + \Big(UX^{\dagger}\tilde{X} - \tilde{U} - UX^{\dagger}\tilde{X}X^{\dagger}X\\
&\quad +\tilde{U}X^{\dagger}X \Big) U^T\tilde{X}\\
&= UX^{\dagger}\tilde{X} + \left(UX^{\dagger}\tilde{X} - \tilde{U} - UX^{\dagger}\tilde{X}X^{\dagger}X + \tilde{U}X^{\dagger}X\right)\\
&\quad \times \left(X^T\tilde{U} - I\right)\\
&= UX^{\dagger}\tilde{X} + \left(UX^{\dagger}\tilde{X} + \tilde{Q}^T - \tilde{U}\right)\left(X^T\tilde{U} - I\right)\\
&= UX^{\dagger}\tilde{X} - UX^{\dagger}\tilde{X} - \tilde{Q}^T + \tilde{U}\\
&\quad + \left(UX^{\dagger}\tilde{X} + \tilde{Q}^T - \tilde{U}\right)X^T\tilde{U},\\
Q\tilde{X} + \tilde{Q}^T &= \tilde{U} + \left(UX^{\dagger}\tilde{X} + \tilde{Q}^T - \tilde{U}\right)X^T\tilde{U}\\
&= \tilde{U} + \left(UX^{\dagger}\tilde{X} + \tilde{U}X^{\dagger}X - UX^{\dagger}\tilde{X}X^{\dagger}X - \tilde{U}\right)X^T\tilde{U}\\
&= \tilde{U} + \left(UX^{\dagger}\tilde{X}\left(I - X^{\dagger}X\right)X^T + \tilde{U}\left(X^{\dagger}X - I\right)X^T\right)\tilde{U}\\
&= \tilde{U}.
\end{aligned}$$

Consequently,

$$Q^*X^* = \begin{pmatrix} X^{\dagger}X & O \\ QX & \tilde{U} \end{pmatrix}.$$

Note that

$$\begin{aligned} 0 &= X^{\dagger}\tilde{X}X^{\dagger}X - X^{\dagger}\tilde{X}X^{\dagger}X \\ &= X^{\dagger}\tilde{X}X^{\dagger}X - X^{\dagger}XX^{\dagger}\tilde{X}X^{\dagger}X \\ &= X^{\dagger}\left(\tilde{X} - XX^{\dagger}\tilde{X}\right)X^{\dagger}X, \end{aligned}$$

whereupon

$$\tilde{X} - XX^{\dagger}\tilde{X} = O.$$

Now we will prove that

$$(X^*)^{\dagger} = \begin{pmatrix} -X^T\tilde{X} & X^{\dagger} \\ I & O \end{pmatrix}.$$

We have

$$\begin{aligned} X^*\begin{pmatrix} -X^{\dagger}\tilde{X} & X^{\dagger} \\ I & O \end{pmatrix}X^* &= \begin{pmatrix} O & I \\ X & \tilde{X} \end{pmatrix}\begin{pmatrix} -X^{\dagger}\tilde{X} & X^{\dagger} \\ I & O \end{pmatrix}\begin{pmatrix} O & I \\ X & \tilde{X} \end{pmatrix} \\ &= \begin{pmatrix} O & I \\ X & \tilde{X} \end{pmatrix}\begin{pmatrix} X^{\dagger}X & O \\ O & I \end{pmatrix} \\ &= \begin{pmatrix} O & I \\ XX^{\dagger}X & \tilde{X} \end{pmatrix} \\ &= \begin{pmatrix} O & I \\ X & \tilde{X} \end{pmatrix} \\ &= X^*, \end{aligned}$$

$$\begin{aligned} &\begin{pmatrix} -X^{\dagger}\tilde{X} & X^{\dagger} \\ I & O \end{pmatrix}X^*\begin{pmatrix} -X^{\dagger}\tilde{X} & X^{\dagger} \\ I & O \end{pmatrix} \\ &= \begin{pmatrix} -X^{\dagger}\tilde{X} & X^{\dagger} \\ I & O \end{pmatrix}\begin{pmatrix} O & I \\ X & \tilde{X} \end{pmatrix}\begin{pmatrix} -X^{\dagger}\tilde{X} & X^{\dagger} \\ I & O \end{pmatrix} \\ &= \begin{pmatrix} -X^{\dagger}\tilde{X} & X^{\dagger} \\ I & O \end{pmatrix}\begin{pmatrix} I & O \\ -XX^{\dagger}\tilde{X} + \tilde{X} & XX^{\dagger} \end{pmatrix} \\ &= \begin{pmatrix} -X^{\dagger}\tilde{X} & X^{\dagger} \\ I & O \end{pmatrix}\begin{pmatrix} I & O \\ O & XX^{\dagger} \end{pmatrix} \end{aligned}$$

$$= \begin{pmatrix} -X^{\dagger}\tilde{X} & X^{\dagger}XX^{\dagger} \\ I & O \end{pmatrix}$$

$$= \begin{pmatrix} -X^{\dagger}\tilde{X} & X^{\dagger} \\ I & O \end{pmatrix}.$$

Note that

$$X^{*}\begin{pmatrix} -X^{\dagger}\tilde{X} & X^{\dagger} \\ I & O \end{pmatrix} = \begin{pmatrix} O & I \\ X & \tilde{X} \end{pmatrix}\begin{pmatrix} -X^{\dagger}\tilde{X} & X^{\dagger} \\ I & O \end{pmatrix}$$

$$= \begin{pmatrix} I & O \\ -XX^{\dagger}\tilde{X}+\tilde{X} & XX^{\dagger} \end{pmatrix}$$

$$= \begin{pmatrix} I & O \\ O & XX^{\dagger} \end{pmatrix},$$

$$\begin{pmatrix} -X^{\dagger}\tilde{X} & X^{\dagger} \\ I & O \end{pmatrix}X^{*} = \begin{pmatrix} -X^{\dagger}\tilde{X} & X^{\dagger} \\ I & O \end{pmatrix}\begin{pmatrix} O & I \\ X & \tilde{X} \end{pmatrix}$$

$$= \begin{pmatrix} X^{\dagger}X & O \\ O & I \end{pmatrix}.$$

Therefore,

$$X^{*}\begin{pmatrix} -X^{\dagger}\tilde{X} & X^{\dagger} \\ I & O \end{pmatrix} \quad \text{and} \quad \begin{pmatrix} -X^{\dagger}\tilde{X} & X^{\dagger} \\ I & O \end{pmatrix}X^{*}$$

are symmetric. Consequently,

$$(X^{*})^{\dagger} = \begin{pmatrix} -X^{\dagger}\tilde{X} & X^{\dagger} \\ I & O \end{pmatrix}.$$

Hence,

$$U^{*}\,(X^{*})^{\dagger}X^{*} = \begin{pmatrix} I & O \\ U & \tilde{U} \end{pmatrix}\begin{pmatrix} X^{\dagger}X & O \\ O & I \end{pmatrix}$$

$$= \begin{pmatrix} X^{\dagger}X & O \\ UX^{\dagger}X & \tilde{U} \end{pmatrix}$$

$$= \begin{pmatrix} X^{\dagger}X & O \\ QX & \tilde{U} \end{pmatrix}.$$

From here,

$$Q^{*}X^{*} = U^{*}\,(X^{*})^{\dagger}X^{*}.$$

Moreover, Q^{*} is differentiable on $[a,b]$ and $\tilde{Q} = X^{-1}$ at dense points of $[a,b]$.

Theorem 4.14 *Let* (X,U) *and* $(\tilde{X},\tilde{U})$ *be normalized conjoined bases of* (4.6). *Then*

$$Ker(X^*)^\beta \subseteq KerX^*$$

if and only if

$$KerX^\beta \subseteq KerX.$$

Proof 1. Let

$$Ker(X^*)^\beta \subseteq KerX^*.$$

Suppose that $x \in KerX^\beta$. Then $\begin{pmatrix} x \\ 0 \end{pmatrix} \in Ker(X^*)^\beta$. Hence, $\begin{pmatrix} x \\ 0 \end{pmatrix} \in KerX^*$ and $x \in KerX$. Because $x \in KerX^\beta$ was arbitrarily chosen and for it we get that it is an element of $KerX$, we conclude that

$$KerX^\beta \subseteq KerX.$$

2. Let

$$KerX^\beta \subseteq KerX.$$

Suppose that $\begin{pmatrix} x \\ 0 \end{pmatrix} \in Ker(X^*)^\beta$. Then $x \in KerX^\beta$ and hence, $x \in KerX$. Therefore, $\begin{pmatrix} x \\ 0 \end{pmatrix} \in KerX^*$ and

$$Ker(X^*)^\beta \subseteq KerX^*.$$

This completes the proof. □

Let

$$D^* = X^*(X^{*\beta})^\dagger \tilde{A}^* B^*.$$

We have

$$(X^*)^{\dagger\beta} = \begin{pmatrix} -(X^\beta)^\dagger \tilde{X}^\beta & (X^\beta)^\dagger \\ I & O \end{pmatrix},$$

$$X^*(X^{*\beta})^\dagger = \begin{pmatrix} O & I \\ X & \tilde{X} \end{pmatrix} \begin{pmatrix} -(X^\beta)^\dagger \tilde{X}^\beta & (X^\beta)^\dagger \\ I & O \end{pmatrix}$$

$$= \begin{pmatrix} I & O \\ -X(X^\beta)^\dagger \tilde{X}^\beta + \tilde{X} & X(X^\beta)^\dagger \end{pmatrix},$$

$$\tilde{A}^* B^* = \begin{pmatrix} I & O \\ O & \tilde{A} \end{pmatrix} \begin{pmatrix} O & O \\ O & B \end{pmatrix}$$

$$= \begin{pmatrix} O & O \\ O & \tilde{A}B \end{pmatrix},$$

$$\begin{aligned}
X^*(X^{*\beta})^{\dagger}\tilde{A}^*B^* &= \begin{pmatrix} I & O \\ -X(X^{\beta})^{\dagger}\tilde{X}^{\beta}+\tilde{X} & X(X^{\beta})^{\dagger} \end{pmatrix}\begin{pmatrix} O & O \\ O & \tilde{A}B \end{pmatrix} \\
&= \begin{pmatrix} O & O \\ O & X(X^{\beta})^{\dagger}\tilde{A}B \end{pmatrix} \\
&= \begin{pmatrix} O & O \\ O & D \end{pmatrix},
\end{aligned}$$

i.e.,

$$D^* = \begin{pmatrix} O & O \\ O & D \end{pmatrix}.$$

Also,

$$\begin{aligned}
(Q^*)^{\beta} &= \begin{pmatrix} -(X^{\beta})^{\dagger}\tilde{X}^{\beta}(X^{\beta})^{\dagger}X^{\beta} & \tilde{Q}^{\beta} \\ (\tilde{Q}^{\beta})^T & Q^{\beta} \end{pmatrix}, \\
(\beta - I)C^* &= \begin{pmatrix} O & O \\ O & (\beta - I)C \end{pmatrix}, \\
(Q^*)^{\beta} + (\beta - I)C^* &= \begin{pmatrix} -(X^{\beta})^{\dagger}\tilde{X}^{\beta}(X^{\beta})^{\dagger}X^{\beta} & \tilde{Q}^{\beta} \\ (\tilde{Q}^{\beta})^T & Q^{\beta} \end{pmatrix} \\
&\quad + \begin{pmatrix} O & O \\ O & (\beta - I)C \end{pmatrix} \\
&= \begin{pmatrix} -(X^{\beta})^{\dagger}\tilde{X}^{\beta}(X^{\beta})^{\dagger}X^{\beta} & \tilde{Q}^{\beta} \\ (\tilde{Q}^{\beta})^T & (\beta - I)C + Q^{\beta} \end{pmatrix}, \\
&\left((Q^*)^{\beta} + (\beta - I)C^*\right)\tilde{A}^*B^* \\
&= \begin{pmatrix} -(X^{\beta})^{\dagger}\tilde{X}^{\beta}(X^{\beta})^{\dagger}X^{\beta} & \tilde{Q}^{\beta} \\ (\tilde{Q}^{\beta})^T & (\beta - I)C + Q^{\beta} \end{pmatrix}\begin{pmatrix} O & O \\ O & \tilde{A}B \end{pmatrix} \\
&= \begin{pmatrix} O & \tilde{Q}^{\beta}\tilde{A}B \\ O & \left((\beta - I)C + Q^{\beta}\right)\tilde{Q}^{\beta}\tilde{A}B \end{pmatrix}, \\
(\tilde{A}^*)^T &= \begin{pmatrix} I & O \\ O & \tilde{A}^T \end{pmatrix}, \\
B^*(\tilde{A}^*)^T &= \begin{pmatrix} O & O \\ O & B \end{pmatrix}\begin{pmatrix} I & O \\ O & \tilde{A}^T \end{pmatrix} \\
&= \begin{pmatrix} O & O \\ O & B\tilde{A}^T \end{pmatrix},
\end{aligned}$$

$$\begin{aligned}
&(\beta - I)B^* \left(\tilde{A}^*\right)^T \left((Q^*)^\beta + (\beta - I)C^*\right) \tilde{A}^* B^* \\
&= (\beta - I) \begin{pmatrix} O & O \\ O & B\tilde{A}^T \end{pmatrix} \begin{pmatrix} O & \tilde{Q}^\beta \tilde{A} B \\ O & \left((\beta - I)C + Q^\beta\right) \tilde{Q}^\beta \tilde{A} B \end{pmatrix} \\
&= (\beta - I) \begin{pmatrix} O & O \\ O & B\tilde{A}^T ((\beta - I)C + Q^\beta) \tilde{Q}^\beta \tilde{A} B \end{pmatrix} \\
&= \begin{pmatrix} O & O \\ O & (\beta - I)B\tilde{A}^T ((\beta - I)C + Q^\beta) \tilde{Q}^\beta \tilde{A} B \end{pmatrix}, \\
&B^* - (\beta - I)B^* \left(\tilde{A}^*\right)^T \left((Q^*)^\beta + (\beta - I)C^*\right) \tilde{A}^* B^* \\
&= \begin{pmatrix} O & O \\ O & B - (\beta - I)B\tilde{A}^T ((\beta - I)C + Q^\beta) \tilde{Q}^\beta \tilde{A} B \end{pmatrix} \\
&= \begin{pmatrix} O & O \\ O & D \end{pmatrix} \\
&= D^*.
\end{aligned}$$

Now we define

$$x^* = \begin{pmatrix} \alpha \\ x \end{pmatrix}, \quad u^* = \begin{pmatrix} 0 \\ u \end{pmatrix} \quad on \quad [a, b].$$

Theorem 4.15 *(x^*, u^*) is admissible if and only if (x, u) is admissible.*

Proof We have

$$\begin{aligned}
B^* U^* &= \begin{pmatrix} O & O \\ O & B \end{pmatrix} \begin{pmatrix} I & O \\ U & \tilde{U} \end{pmatrix} \\
&= \begin{pmatrix} O & O \\ BU & B\tilde{U} \end{pmatrix}, \\
(X^*)^\beta &= \begin{pmatrix} O & I \\ X^\beta & \tilde{X}^\beta \end{pmatrix}, \\
A^* (X^*)^\beta &= \begin{pmatrix} O & O \\ O & A \end{pmatrix} \begin{pmatrix} O & I \\ X^\beta & \tilde{X}^\beta \end{pmatrix} \\
&= \begin{pmatrix} O & O \\ AX^\beta & A\tilde{X}^\beta \end{pmatrix}.
\end{aligned}$$

Then

$$D_\beta(X^*) = \begin{pmatrix} O & O \\ D_\beta X & D_\beta \tilde{X} \end{pmatrix}$$

$$= A^*(X^*)^\beta + B^* U^*$$

$$= \begin{pmatrix} O & O \\ AX^\beta & A\tilde{X}^\beta \end{pmatrix} + \begin{pmatrix} O & O \\ BU & B\tilde{U} \end{pmatrix}$$

$$= \begin{pmatrix} O & O \\ AX^\beta + BU & A\tilde{X}^\beta + B\tilde{U} \end{pmatrix}.$$

Hence, (x^*, u^*) is admissible if and only if (x, u) is admissible. This completes the proof. □

Exercise 4.2 Let (x^*, u^*) be admissible. Prove that

$$x^* \in ImX^*$$

if and only if

$$\alpha + U^T x \in ImX^T \quad on \quad [a, b].$$

Let

$$Q_1^* = \begin{pmatrix} O & O \\ O & Q \end{pmatrix}.$$

Theorem 4.16 (The Extended Picone Identity)

Suppose that hypothesis (A) holds. Let be given an admissible (x, u) and a constant $\alpha \in \mathbb{R}^n$ with $\alpha + U^T(a)x(a) \in ImX^T(a)$. Then $\alpha + U^T x \in ImX^T$ on $[a, b]$,

$$\mathscr{F}_0(x, u) = \begin{pmatrix} \alpha \\ x(b) \end{pmatrix}^T Q_1^*(b) \begin{pmatrix} \alpha \\ x(b) \end{pmatrix} - \begin{pmatrix} \alpha \\ x(a) \end{pmatrix}^T Q_1^*(a) \begin{pmatrix} \alpha \\ x(a) \end{pmatrix}$$

$$+ \int_a^b (z^T D z)(t) D_\beta t$$

and

$$x + (\beta - I)Dz = \left(\tilde{A}^{-1} - (\beta - I)B\tilde{A}^T(Q^\beta + (\beta - I)C)\right)x^\beta - (\beta - I)B\tilde{A}^T(\tilde{Q}^\beta)^T\alpha$$

on $[a, b]$, where

$$z = u - Qx - \tilde{Q}^T\alpha \quad on \quad [a, b].$$

Proof Since (x, u) is admissible, we have that (x^*, u^*) is admissible. Therefore, $\alpha + U^T x \in ImX^T$ and from here, $x^* \in ImX^*$ on $[a, b]$. Let

$$z^* = u^* - Q^* x^*.$$

We have

$$
\begin{aligned}
z^* &= \begin{pmatrix} 0 \\ u \end{pmatrix} - \begin{pmatrix} -X^\dagger \tilde{X} X^\dagger X & \tilde{Q} \\ \tilde{Q}^T & Q \end{pmatrix} \begin{pmatrix} \alpha \\ x \end{pmatrix} \\
&= \begin{pmatrix} X^\dagger \tilde{X} X^\dagger X \alpha - \tilde{Q} x \\ u - \tilde{Q}^T \alpha - Q x \end{pmatrix} \\
&= \begin{pmatrix} X^\dagger \tilde{X} X^\dagger X \alpha - \tilde{Q} x \\ z \end{pmatrix}.
\end{aligned}
$$

Then

$$
\begin{aligned}
(u^*)^T B^* u^* &= (0, u^T) \begin{pmatrix} 0 & 0 \\ 0 & B \end{pmatrix} \begin{pmatrix} 0 \\ u \end{pmatrix} \\
&= (0, u^T) \begin{pmatrix} 0 \\ Bu \end{pmatrix} \\
&= u^T B u,
\end{aligned}
$$

$$
\begin{aligned}
\left((x^*)^\beta\right)^T C^* (x^*)^\beta &= (\alpha^T, (x^\beta)^T) \begin{pmatrix} 0 & 0 \\ 0 & C \end{pmatrix} \begin{pmatrix} \alpha \\ x^\beta \end{pmatrix} \\
&= (\alpha^T, (x^\beta)^T) \begin{pmatrix} 0 \\ C x^\beta \end{pmatrix} \\
&= (x^\beta)^T C x^\beta,
\end{aligned}
$$

whereupon

$$
u^T B u - (x^\beta)^T C x^\beta = (u^*)^T B^* u^* - \left((x^*)^\beta\right)^T C^* (x^*)^\beta.
$$

Also,

$$
\begin{aligned}
(x^*)^T Q_1^* x^* &= (\alpha^T, x^T) \begin{pmatrix} O & O \\ O & Q \end{pmatrix} \begin{pmatrix} \alpha \\ x \end{pmatrix} \\
&= (\alpha^T, x^T) \begin{pmatrix} 0 \\ Qx \end{pmatrix} \\
&= x^T Q x,
\end{aligned}
$$

$$(z^*)^T Dz^* = \left((X^\dagger \tilde{X} X^\dagger X\alpha)^T - x^T \tilde{Q}^T, z^T \right) \times \begin{pmatrix} O & O \\ O & D \end{pmatrix} \begin{pmatrix} X^\dagger \tilde{X} X^\dagger X\alpha - \tilde{Q}x \\ z \end{pmatrix} = z^T Dz.$$

Therefore,

$$\begin{aligned}
\mathscr{F}_0(x,u) &= \int_a^b \left(u^T Bu - (x^\beta)^T Cx^\beta \right)(t) D_\beta t \\
&= \mathscr{F}_0(x^*, u^*) \\
&= \int_a^b \left(D_\beta \left((x^*)^T Q_1^* x^* \right) + (z^*)^T D^* z^* \right)(t) D_\beta t \\
&= (x^*)^T(b) Q_1^*(b) x^*(b) - (x^*)^T(a) Q_1^*(a) x^*(a) \\
&\quad + \int_a^b \left((z^*)^T D^* z^* \right)(t) D_\beta t \\
&= \begin{pmatrix} \alpha \\ x(b) \end{pmatrix}^T Q_1^*(b) \begin{pmatrix} \alpha \\ x(b) \end{pmatrix} - \begin{pmatrix} \alpha \\ x(a) \end{pmatrix}^T Q_1^*(a) \begin{pmatrix} \alpha \\ x(a) \end{pmatrix} \\
&\quad + \int_a^b \left(z^T Dz \right)(t) D_\beta t.
\end{aligned}$$

Note that

$$\begin{aligned}
x^* + (\beta - I) D^* z^* &= \begin{pmatrix} \alpha \\ x \end{pmatrix} + (\beta - I) \begin{pmatrix} O & O \\ O & D \end{pmatrix} \begin{pmatrix} X^\dagger \tilde{X} X^\dagger X\alpha - \tilde{Q}x \\ z \end{pmatrix} \\
&= \begin{pmatrix} \alpha \\ x \end{pmatrix} + (\beta - I) \begin{pmatrix} 0 \\ Dz \end{pmatrix} \\
&= \begin{pmatrix} \alpha \\ x + (\beta - I) Dz \end{pmatrix}.
\end{aligned}$$

Hence, applying Theorem 4.12, we get

$$x^* + (\beta - I)D^* z^* = \begin{pmatrix} \alpha \\ \left(\tilde{A}^{-1} - (\beta - I)B\tilde{A}^T \left(Q^\beta + (\beta - I)C\right)\right) x^\beta \end{pmatrix}. \tag{4.20}$$

Observe that

$$\begin{aligned}
&(\tilde{A}^*)^{-1} - (\beta - I)B^*(\tilde{A}^*)^T \left((Q^*)^\beta + (\beta - I)C^*\right) \\
&= \begin{pmatrix} I & O \\ O & \tilde{A}^{-1} \end{pmatrix} - (\beta - I)\begin{pmatrix} O & O \\ O & B \end{pmatrix}\begin{pmatrix} I & O \\ O & \tilde{A}^T \end{pmatrix} \\
&\quad \times \left(\begin{pmatrix} -\left(X^\dagger \tilde{X} X^\dagger X\right)^\beta & \tilde{Q}^\beta \\ \tilde{Q}^{\beta T} & Q^\beta \end{pmatrix} + (\beta - I)\begin{pmatrix} O & O \\ O & C \end{pmatrix}\right) \\
&= \begin{pmatrix} I & O \\ O & \tilde{A}^{-1} \end{pmatrix} - (\beta - I)\begin{pmatrix} O & O \\ O & B\tilde{A}^T \end{pmatrix} \\
&\quad \times \begin{pmatrix} -\left(X^\dagger \tilde{X} X^\dagger X\right)^\beta & \tilde{Q}^\beta \\ \tilde{Q}^{\beta T} & Q^\beta + (\beta - I)C \end{pmatrix} \\
&= \begin{pmatrix} I & O \\ O & \tilde{A}^{-1} \end{pmatrix} - (\beta - I)\begin{pmatrix} O & O \\ B\tilde{A}^T \tilde{Q}^{\beta T} & B\tilde{A}^T \left(Q^\beta + (\beta - I)C\right) \end{pmatrix} \\
&= \begin{pmatrix} I & O \\ -(\beta - I)B\tilde{A}^T \tilde{Q}^{\beta T} & \tilde{A}^{-1} - (\beta - I)B\tilde{A}^T \left(Q^\beta + (\beta - I)C\right) \end{pmatrix},
\end{aligned}$$

$$\begin{aligned}
&\left((\tilde{A}^*)^{-1} - (\beta - I)B^*(\tilde{A}^*)^T \left((Q^*)^\beta + (\beta - I)C^*\right)\right)(x^*)^\beta \\
&= \begin{pmatrix} I & O \\ -(\beta - I)B\tilde{A}^T \tilde{Q}^{\beta T} & \tilde{A}^{-1} - (\beta - I)B\tilde{A}^T \left(Q^\beta + (\beta - I)C\right) \end{pmatrix}\begin{pmatrix} \alpha \\ x^\beta \end{pmatrix} \\
&= \begin{pmatrix} \alpha \\ \left(\tilde{A}^{-1} - (\beta - I)B\tilde{A}^T \left(Q^\beta + (\beta - I)C\right)\right) x^\beta - (\beta - I)B\tilde{A}^T \tilde{Q}^{\beta T} \alpha \end{pmatrix}.
\end{aligned}$$

From the last equality and from (4.20), we obtain

$$\begin{aligned}
&\left(\left(\tilde{A}^*\right)^{-1} - (\beta - I)B^* \left(\tilde{A}^*\right)^T \left((Q^*)^\beta + (\beta - I)C^*\right)\right)(x^*)^\beta \\
&\qquad = x^* + (\beta - I)D^* z^*.
\end{aligned}$$

This completes the proof. □

4.7 Positivity of Quadratic Functionals

Theorem 4.17 (Sufficient Condition for Positive Definiteness)
Suppose that there exists a conjoined basis of the system (4.6) *with no focal point in* $(a,b]$. *Then the quadratic functional* $\mathscr{F}_0$ *is positive definite.*

Proof Suppose that (X,U) is a conjoined basis of the system (4.6) such that X is nonsingular at right-dense points of $[a,b]$ and

$$KerX^{\beta} \subseteq KerX, \quad D = X(X^{\beta})^{\dagger}\tilde{A}B \geq 0.$$

By Theorem 4.9, it follows that there exists another conjoined basis $(\tilde{X},\tilde{U})$ of (4.6) such that (X,U) and $(\tilde{X},\tilde{U})$ are normalized. Let (x,u) be admissible with $x(a) = x(b) = 0$. Then $x(a) \in ImX(a)$ and by Theorem 4.12 we get that $x(t) \in ImX(t)$ for all $t \in [a,b]$.

1. Suppose that a is right-scattered. Then, using Theorem 4.12, we obtain

$$\mathscr{F}_0(x,u) = \int_a^b \left(u^T Bu - (x^{\beta})^T Cx^{\beta}\right)(t)D_{\beta}t$$

$$= \int_a^{\beta(a)} \left(u^T Bu - (x^{\beta})^T Cx^{\beta}\right)(t)D_{\beta}t$$

$$+ \int_{\beta(a)}^b \left(u^T Bu - (x^{\beta})^T Cx^{\beta}\right)(t)D_{\beta}t$$

$$= (\beta - I)(a)\left(u^T Bu - (x^{\beta})^T Cx^{\beta}\right)(a)$$

$$+(x(b))^T Q(b)x(b) - (x^{\beta}(a))^T Q^{\beta}(a)x^{\beta}(a)$$

$$+ \int_{\beta(a)}^b (z^T Dz)(t)D_{\beta}t$$

$$\geq \left((\beta - I)u^T Bu - (x^{\beta})^T (Q^{\beta} + (\beta - I)C)x^{\beta}\right)(a)$$

$$= \left((\beta - I)u^T Bu - (x + (\beta - I)Bu)^T \tilde{A}^T (Q^{\beta} + (\beta - I)C)\tilde{A}(x + (\beta - I)Bu)\right)(a)$$

$$= \left((\beta - I)u^T (B - (\beta - I)B\tilde{A}^T (Q^\beta + (\beta - I)C)\tilde{A}B)u \right.$$

$$\left. - x^T \tilde{A}^T (Q^\beta + (\beta - I)C)\tilde{A}(x + 2(\beta - I)Bu) \right) (a)$$

$$= \left((\beta - I)z^T Dz \right)(a)$$

$$\geq 0.$$

Now we assume that $\mathscr{F}_0(x,u) = 0$. Then, using Theorem 4.12, we get

$$\int_a^b (z^T Dz)(t) D_\beta t = 0.$$

Since D is positive definite, we obtain

$$z^T Dz = 0 \quad on \quad [\beta(a), b]$$

and

$$Dz = 0 \quad on \quad [\beta(a), b].$$

2. Now we suppose that a is right-dense. We take a sequence $\{a_m\}_{m\in\mathbb{N}} \subseteq (a,b]$ such that $a_m \to a$, as $m \to \in fty$. Since $x \in ImX$ on $[a,b]$, we have

$$(I - X^\dagger X)U^T x = (I - X^\dagger X)U^T Xd$$

$$= (I - X^\dagger X)X^T Ud$$

$$= 0$$

for some $d \in \mathbb{R}^n$. Hence,

$$\tilde{Q}x = X^\dagger x \quad on \quad (a,b]$$

and

$$Qx = UX^\dagger x \quad on \quad (a,b].$$

Also,

$$x^T UX^\dagger \tilde{X} X^\dagger X U^T x = x^T UX^\dagger \tilde{X} (X^\dagger)^T U^T x$$

$$= x^T UX^\dagger X \tilde{X}^T X^T (X^\dagger)^T U^T x$$

$$= x^T QX\tilde{X}^T (X^\dagger)^T U^T x$$

$$= x^T Q^T \tilde{X} X^T (X^\dagger)^T U^T x$$

$$= x^T Q^T \tilde{X} X^\dagger X U^T x.$$

Note that

$$Q^T = (X^\dagger)^T U^T + U(I - X^\dagger X)(X^\dagger \tilde{X} U^T - \tilde{U}^T),$$

$$Q^T \tilde{X} = (X^\dagger)^T U^T \tilde{X} + U(I - X^\dagger X)(X^\dagger \tilde{X} U^T - \tilde{U}^T)\tilde{X}$$

$$= (X^\dagger)^T U^T \tilde{X}$$

$$= (X^\dagger)^T (X^T \tilde{U} - I).$$

Therefore,

$$x^T UX^\dagger \tilde{X} X^\dagger X U^T x = x^T (X^\dagger)^T (X^T \tilde{U} - I) X^\dagger X U^T x$$

$$= x^T (X^\dagger)^T X^T \tilde{U} X^\dagger X U^T x$$

$$- x^T (X^\dagger)^T X^\dagger X U^T x$$

$$= x^T X X^\dagger \tilde{U} X^\dagger X U^T x - x^T (X^\dagger)^T X^T (X^\dagger)^T U^T x$$

$$= x^T X X^\dagger \tilde{U} X^\dagger X U^T x - x^T (X^\dagger)^T U^T x$$

$$= x^T X X^\dagger \tilde{U} X^\dagger X \alpha_m - x^T Qx,$$

where

$$\alpha_m = U^T x(t_m).$$

Now we apply Theorem 4.16 on the interval $[a_m, b]$ with $\alpha = \alpha_m$ and

$$z_m = u - Qx - \tilde{Q}^T \alpha_m \quad on \quad [a,b]$$

and we get

$$\begin{aligned}
\int_{a_m}^{b} (u^T Bu - (x^\beta)^T Cx^\beta)(t) D_\beta t &= \begin{pmatrix} \alpha_m \\ x(b) \end{pmatrix}^T Q^*(b) \begin{pmatrix} \alpha_m \\ x(b) \end{pmatrix} \\
&\quad - \begin{pmatrix} \alpha_m \\ x(a_m) \end{pmatrix}^T Q^*(a) \begin{pmatrix} \alpha_m \\ x(a_m) \end{pmatrix} \\
&\quad + \int_{a_m}^{b} (z_m^T D z_m)(t) D_\beta t \\
&\geq -(\alpha_m^T X^\dagger \tilde{X} X^\dagger X \alpha_m)(b) \\
&\quad - \left(x^T Qx + 2\alpha_m^T \tilde{Q} x - \alpha_m^T X^\dagger \tilde{X} X^\dagger X \alpha_m\right)(a_m) \\
&= -(\alpha_m^T X^\dagger \tilde{X} X^\dagger X \alpha_m)(b) \\
&\quad - \left(x^T Qx - 2x^T U \tilde{Q} x - x^T U X^\dagger \tilde{X} X^\dagger X U^T x\right)(a_m) \\
&= -(\alpha_m^T X^\dagger \tilde{X} X^\dagger X \alpha_m)(b) \\
&\quad - \left(x^T Qx - 2x^T U X^\dagger x - x^T U X^\dagger \tilde{X} X^\dagger X U^T x\right)(a_m) \\
&= -(\alpha_m^T X^\dagger \tilde{X} X^\dagger X \alpha_m)(b) \\
&\quad - \left(x^T Qx - 2x^T Qx - x^T U X^\dagger \tilde{X} X^\dagger X U^T x\right)(a_m) \\
&= -(\alpha_m^T X^\dagger \tilde{X} X^\dagger X \alpha_m)(b) \\
&\quad - \left(-x^T Q^* - x^T U X^\dagger \tilde{X} X^\dagger X U^T x\right)(a_m) \\
&= -(\alpha_m^T X^\dagger \tilde{X} X^\dagger X \alpha_m)(b) \\
&\quad + \left(x^T U X^\dagger \tilde{X} X^\dagger X \alpha_m\right)(a_m).
\end{aligned}$$

Now, using that $\alpha_m \to 0$, as $m \to \in fty$, we get

$$\begin{aligned}
\mathscr{F}_0(x,u) &= \lim_{m\to\in fty} \int_{a_m}^{b} \left(u^T Bu - (x^\beta)^T C x^\beta\right)(t) D_\beta t \\
&\geq \lim_{m\to\in fty} \Bigg(-(\alpha_m^T X^\dagger \tilde{X} X^\dagger X \alpha_m)(b) \\
&\quad (x^T U X^\dagger \tilde{X} X^\dagger X \alpha_m)(a_m) \Bigg) \\
&= 0.
\end{aligned}$$

If $\mathscr{F}_0(x,u) = 0$, using

$$\lim_{m\to\in fty} \int_{a_m}^{b} (z_m^T D z_m)(t) D_\beta t = 0,$$

where

$$z_m \to z = u - Qx, \quad as \quad m \to \in fty,$$

holds uniformly on $[a_k, b]$ for any $k \in \mathbb{N}$. We fix $k \in \mathbb{N}$. Observe that

$$\int_{a_m}^{b} (z_m^T D z_m)(t) D_\beta t \geq \int_{a_k}^{b} (z_m^T D z_m)(t) D_\beta t$$

for any $m \geq k$. Therefore,

$$\lim_{m\to\in fty} \int_{a_m}^{b} (z_m^T D z_m)(t) D_\beta t \geq \int_{a_k}^{b} (z^T D z)(t) D_\beta t$$

and hence,

$$\begin{aligned}
0 &= \lim_{m\to\in fty} \int_{a_m}^{b} (z_m^T D z_m)(t) D_\beta t \\
&\geq \int_{a}^{b} (z^T D z)(t) D_\beta t \\
&\geq 0.
\end{aligned}$$

Consequently,

$$z^T Dz = 0 \quad on \quad [a,b].$$

Therefore,

$$Dz = 0 \quad on \quad [\beta(a), b].$$

We have

$$x = \left(\tilde{A}^{-1} - (\beta - I)B\tilde{A}^T(Q^\beta + (\beta - I)C)\right)x^\beta \quad on \quad [\beta(a), b]. \tag{4.21}$$

Now we will show that $x = 0$ on $[a,b]$. Consider the statement

$$L(t) = \{x(t) = 0 \quad on \quad [\beta(a), b]\}.$$

1. We have that $L(b)$ is true because $x(b) = 0$.
2. Let $t < \beta(t)$ and $L(t)$ be true. Then $\beta^{-1}(t)$ is right-scattered and using (4.21), we get $L(\beta^{-1}(t))$ is true.
3. Let $t \in (\beta(a), b], t = \beta(t)$ and $L(t)$ be true. Since X is nonsingular at t, there exists a neighborhood V of t such that X is nonsingular on V. Because $KerX^\beta(t) \subseteq KerX(t)$, we get that $X^\beta(t)$ is nonsingular. Therefore,

$$Dz = 0 \quad \Longleftrightarrow \quad Bu = BUX^{-1}x \quad on \quad V,$$

and we obtain

$$D_\beta x = \tilde{A}(A + BUX^{-1})x \quad on \quad V.$$

Note that

$$\begin{aligned} I + (\beta - I)\tilde{A}(A + BUX^{-1}) &= \tilde{A}(I - (\beta - I)A + (\beta - I)A + (\beta - I)BUX^{-1}) \\ &= \tilde{A}(I + (\beta - I)BUX^{-1}) \\ &= \tilde{A}(X + (\beta - I)BU)X^{-1} \\ &= X^\beta X^{-1}. \end{aligned}$$

Therefore,

$$\tilde{A}(A + BUX^{-1})$$

is regressive on V. Hence, the IVP

$$\begin{cases} D_\beta x = \tilde{A}(\tau)(A(\tau) + B(\tau)U(\tau)X^{-1}(\tau))x, & \tau \in V, \\ x(t) = 0 \end{cases}$$

has unique solution $x = 0$ on V. Therefore, $L(s)$ is true for $s \in V$, $s < t$.

4. Let $t \in [\beta(a), b)$ be right-dense and $L(s)$ is true for $s \in (t, b]$. Since x is continuous at t, we have

$$x(t) = \lim_{s \to t} x(s)$$

$$= 0,$$

i.e., $L(t)$ holds.

Therefore, $L(t)$ is true for all $t \in [\beta(a), b]$ and since $x(a) = 0$, we have $x = 0$ on $[a, b]$. Consequently, $\mathscr{F}_0 > 0$. This completes the proof. □

Chapter 5
The First Variation

Let $I \subset \mathbb{R}$ and β be a general first kind or second kind quantum operator.

5.1 The Dubois-Reymond Lemma

In this section several versions of the fundamental lemma of Dubois-Reymond of general quantum variational calculus will be deducted.

Lemma 5.1 *Let $g \in \mathscr{C}_\beta([a,b])$, $g : [a,b] \to \mathbb{R}^n$. Then*

$$\int_a^b g^T(t)\eta(t)D_\beta t = 0 \quad \text{for all} \quad \eta \in \mathscr{C}^1_\beta([a,b]), \quad \eta : [a,b] \to \mathbb{R}^n, \quad \eta(a) = \eta(b) = 0, \tag{5.1}$$

holds if and only if

$$g(t) \equiv 0 \quad on \quad [a,b]. \tag{5.2}$$

Proof 1. Let (5.2) holds. Then it follows (5.1).
2. Suppose that (5.1) holds. Assume that $g(t_1) \neq 0$ for some $t_1 \in [a,b]$. Without loss of generality, we suppose that $g(t_1) > 0$. Since $g \in \mathscr{C}_\beta([a,b])$, there exists $[t_2,t_3] \subset [a,b]$ such that $t_1 \in [t_2,t_3]$ and $g(t) > 0$ for $t \in [t_2,t_3]$. Consider the function

$$\eta(t) = \begin{cases} \begin{pmatrix} (t-t_2)(t_3-t) \\ \vdots \\ (t-t_2)(t_3-t) \end{pmatrix} & for \quad t \in [t_2,t_3] \\ \begin{pmatrix} 0 \\ \vdots \\ 0 \end{pmatrix} & for \quad [a,b] \backslash [t_2,t_3]. \end{cases}$$

Then

$$0 = \int_a^b g^T(t)\eta(t)D_\beta t$$

$$= \sum_{i=1}^n \int_{t_2}^{t_3} g_i(t)(t-t_2)(t_3-t)D_\beta t$$

$$> 0.$$

This is a contradiction. Consequently, $g(t) \equiv 0$ on $[a,b]$. This completes the proof. □

Lemma 5.2 (Dubois-Reymond Lemma)
Let $g \in \mathscr{C}_\beta([a,b]$, $g : [a,b] \to \mathbb{R}^n$. *Then*

$$\int_a^b g^T(t)D_\beta\eta(t)D_\beta t = 0 \quad for \quad all \quad \eta \in \mathscr{C}_\beta^1([a,b]), \quad \eta : [a,b] \to \mathbb{R}^n, \quad \eta(a) = \eta(b) = 0, \tag{5.3}$$

holds if and only if

$$g(t) \equiv c \quad on \quad [a,b] \quad for \quad some \quad c \in \mathbb{R}^n. \tag{5.4}$$

Proof 1. Let (5.4) hold. Then

$$\int_a^b g^T(t)D_\beta\eta(t)D_\beta t = \int_a^b c^T D_\beta\eta(t)D_\beta t$$

$$= c^T \int_a^b D_\beta\eta(t)D_\beta t$$

$$= c^T(\eta(b) - \eta(a))$$

$$= 0.$$

2. Let (5.3) hold. We set

$$G(t) = \int_a^t g(\tau)D_\beta\tau, \quad t \in [a,b], \quad c = \frac{G(b)}{b-a}.$$

Let

$$\eta(t) = G(t) - (t-a)c, \quad t \in [a,b].$$

Then

$$\begin{aligned}
D_\beta \eta(t) &= D_\beta G(t) - c \\
&= g(t) - c, \\
\eta(a) &= G(a) \\
&= 0, \\
\eta(b) &= G(b) - (b-a)c \\
&= G(b) - G(b) \\
&= 0.
\end{aligned}$$

Hence,

$$\begin{aligned}
0 &= \int_a^b g^T(t) D_\beta \eta(t) D_\beta t \\
&= \int_a^b g^T(t) (g(t) - c) D_\beta t \\
&= \int_a^b (g(t) - c)^T (g(t) - c) D_\beta t + c^T \int_a^b (g(t) - c) D_\beta t \\
&= \int_a^b |g(t) - c|^2 D_\beta t + c^T \int_a^b g(t) D_\beta t - c^T c(b-a) \\
&= \int_a^b |g(t) - c|^2 D_\beta t + c^T G(b) - c^T c(b-a) \\
&= \int_a^b |g(t) - c|^2 D_\beta t + c^T (G(b) - c(b-a)) \\
&= \int_a^b |g(t) - c|^2 D_\beta t.
\end{aligned}$$

Therefore, $g(t) \equiv c$ on $[a, b]$. This completes the proof. □

Lemma 5.3 *Let $g \in \mathscr{C}_\beta([a,b])$, $g : [a,b] \to \mathbb{R}^n$. Then*

$$\int_a^b g^T(t) D_\beta^2 \eta(t) D_\beta t = 0 \tag{5.5}$$

for all $\eta \in \mathscr{C}_\beta^2([a,b])$, $\eta : [a,b] \to \mathbb{R}^n$, $\eta(a) = \eta(b) = 0$, $D_\beta \eta(a) = D_\beta \eta(b) = 0$, if and only if

$$g(t) = c_0 + c_1 \beta(t), \quad t \in [a,b], \tag{5.6}$$

for some $c_0, c_1 \in \mathbb{R}^n$.

Proof
Let (5.6) hold. Suppose that $\eta \in \mathscr{C}_\beta^2([a,b])$,$\eta : [a,b] \to \mathbb{R}^n$, $\eta(a) = \eta(b) = 0$, $D_\beta \eta(a) = D_\beta \eta(b) = 0$. Then

$$\begin{aligned}
\int_a^b g^T(t) D_\beta^2 \eta(t) D_\beta t &= \int_a^b \left(c_0^T + c_1^T \beta(t)\right) D_\beta^2 \eta(t) D_\beta t \\
&= \left(c_0^T + c_1^T t\right) D_\beta \eta(t)\Big|_{t=a}^{t=b} - \int_a^b c_1^T D_\beta \eta(t) D_\beta t \\
&= -c_1^T \eta(t)\Big|_{t=a}^{t=b} \\
&= 0,
\end{aligned}$$

i.e., (5.5) holds.
Suppose that (5.5) holds. Let $c_0, c_1 \in \mathbb{R}^n$ be such that

$$\begin{cases}
\displaystyle\int_a^b \int_a^s \left(g(\xi) - c_0 - c_1 \beta(\xi)\right) D_\beta \xi D_\beta s = 0 \\
\displaystyle\int_a^b \left(g(\xi) - c_0 - c_1 \beta(\xi)\right) D_\beta \xi = 0.
\end{cases}$$

We take

$$\eta(t) = \int_a^t \int_a^s \left(g(\xi) - c_0 - c_1 \beta(\xi)\right) D_\beta \xi D_\beta s, \quad t \in [a,b].$$

Then $\eta \in \mathscr{C}^2_\beta([a,b])$ and

$$\eta(a) = 0,$$

$$\eta(b) = \int_a^b \int_a^s (g(\xi) - c_0 - c_1\beta(\xi))\, D_\beta \xi D_\beta s$$

$$= 0,$$

$$D_\beta \eta(t) = \int_a^t (g(\xi) - c_0 - c_1\beta(\xi))\, D_\beta \xi, \quad t \in [a,b],$$

$$D_\beta \eta(a) = 0,$$

$$D_\beta \eta(b) = \int_a^b (g(\xi) - c_0 - c_1\beta(\xi))\, D_\beta \xi$$

$$= 0.$$

Hence,

$$\int_a^b (g^T(t) - c_0^T - c_1^T \beta(t))\,(g(t) - c_0 - c_1\beta(t))\, D_\beta t$$

$$= \int_a^b (g^T(t) - c_0^T - c_1^T \beta(t))\, D_\beta^2 \eta(t) D_\beta t$$

$$= \int_a^b g^T(t) D_\beta^2 \eta(t) D_\beta t - c_0^T \int_a^b D_\beta \eta(t) D_\beta t$$

$$-c_1^T \int_a^b \beta(t) D_\beta^2 \eta(t) D_\beta t$$

$$= -c_1^T t D_\beta \eta(t)\Big|_{t=a}^{t=b} + c_1^T \int_a^b D_\beta \eta(t) D_\beta t$$

$$= c_1^T \eta(t)\Big|_{t=a}^{t=b}$$

$$= 0.$$

Therefore,

$$g(t) \equiv c_0 + c_1\beta(t) \quad on \quad [a,b].$$

This completes the proof. □

Lemma 5.4 *Let* $\alpha, \beta \in \mathscr{C}_\beta([a,b])$, $\alpha, \beta : [a,b] \to \mathbb{R}^n$. *Then*

$$\int_a^b \left(\alpha^T(t)h(t) + \beta^{\beta T}(t)D_\beta h(t)\right) D_\beta t = 0 \tag{5.7}$$

for all $h \in \mathscr{C}^1_\beta([a,b])$, $h : [a,b] \to \mathbb{R}^n$, $h(a) = h(b) = 0$, *if and only if*

$$\beta \in \mathscr{C}^1_\beta([a,b]) \quad D_\beta\beta(t) = \alpha(t), \quad t \in [a,b]. \tag{5.8}$$

Proof 1. Let (5.8) hold. Then

$$\begin{aligned}
\int_a^b \left(\alpha^T(t)h(t) + \beta^{\beta T}(t)D_\beta h(t)\right) D_\beta t &= \int_a^b \alpha^T(t)h(t)D_\beta t \\
&\quad + \int_a^b \beta^{\beta T}(t)D_\beta h(t)D_\beta t \\
&= \int_a^b \alpha^T(t)h(t)D_\beta t + \beta^T(t)h(t)\Big|_{t=a}^{t=b} \\
&\quad - \int_a^b \beta^{d_\beta t}(t)h(t)D_\beta t \\
&= \int_a^b \alpha^T(t)h(t)D_\beta t - \int_a^b \alpha^T(t)h(t)D_\beta t \\
&= 0.
\end{aligned}$$

2. Let (5.7) hold. We set

$$A(t) = \int_a^t \alpha(s)D_\beta s, \quad t \in [a,b].$$

Then

$$\int_a^b \alpha^T(t)h(t)D_\beta t = \int_a^b A^{d_\beta t}(t)h(t)D_\beta t$$

$$= A^T(t)h(t)\Big|_{t=a}^{t=b} - \int_a^b A^{\beta T}(t)D_\beta h(t)D_\beta t$$

$$= -\int_a^b A^{\beta T}(t)D_\beta h(t)D_\beta t.$$

Therefore,

$$0 = \int_a^b \left(\alpha^T(t)h(t) + \beta^{\beta T}(t)D_\beta h(t)\right) D_\beta t$$

$$= \int_a^b \left(-A^{\beta T}(t) + \beta^{\beta T}(t)\right) D_\beta h(t)D_\beta t$$

$$= \int_a^b \left(-A^\beta + \beta^\beta\right)^T (t)D_\beta h(t)D_\beta t.$$

From this and, we obtain

$$-A^\beta(t) + \beta^\beta(t) = c^\beta, \quad t \in [a,b],$$

for some $c \in \mathbb{R}^n$. Consequently,

$$\beta(t) = A(t) + c, \quad t \in [a,b],$$

for some $c \in \mathbb{R}^n$. From here, we conclude that $\beta \in \mathscr{C}_\beta^1([a,b])$ and

$$D_\beta \beta(t) = \alpha(t), \quad t \in [a,b].$$

This completes the proof. □

5.2 The Variational Problem

Consider the variational problem

$$\mathscr{L}(y) = \int_a^b L(t, y^\beta(t), D_\beta y(t)) D_\beta t \longrightarrow \min, \quad y(a) = \alpha, y(b) = \beta, \tag{5.9}$$

where $a, b \in I$ with $a < b$, $\alpha, \beta \in \mathbb{R}^n$, $n \in \mathbb{N}$, and $L : I \times \mathbb{R}^{2n} \to \mathbb{R}$ is a $\mathscr{C}^2$-function in the last two variables, $L = L(t, x, v)$. For $f \in \mathscr{C}^1_\beta([a,b])$ we define the norm

$$\|f\| = \max_{t \in [a,b]} |f^\beta(t)| + \max_{t \in [a,b]} |D_\beta f(t)|.$$

Definition 5.1 The function L is called Lagrangian.

Definition 5.2 A function $\hat{y} \in \mathscr{C}^1_\beta([a,b])$ with $\hat{y}(a) = \alpha$, $\hat{y}(b) = \beta$, is called a (weak) local minimum of (5.9) if there exists $\delta > 0$ such that

$$\mathscr{L}(\hat{y}) \le \mathscr{L}(y)$$

for all $y \in \mathscr{C}^1_\beta([a,b])$ with $y(a) = \alpha$, $y(b) = \beta$ and

$$\|y - \hat{y}\| < \delta.$$

If $\mathscr{L}(\hat{y}) < \mathscr{L}(y)$ for all such that $y \ne \hat{y}$ is said to be proper.

Definition 5.3 An $\eta \in \mathscr{C}^1_\beta([a,b])$ is called an admissible variation provided

$$\eta(a) = \eta(b) = 0.$$

For an admissible variation η, we define a function $\Phi : \mathbb{R} \to \mathbb{R}$ as follows

$$\Phi(\varepsilon) = \Phi(\varepsilon, y, \eta)$$

$$= \mathscr{L}(y + \varepsilon\eta), \quad \varepsilon \in \mathbb{R}.$$

Definition 5.4 The first and second variations of the variational problem (5.9) are defined by

$$\mathscr{L}_1(y, \eta) = \Phi'(0) \quad \textit{and} \quad \mathscr{L}_2(y, \eta) = \Phi''(0),$$

respectively.

Theorem 5.1 (Necessary Condition)
If $\hat{y}$ is a local minimum of (5.9), *then*

$$\mathscr{L}_1(\hat{y},\eta)=0 \quad and \quad \mathscr{L}_2(\hat{y},\eta)\geq 0$$

for all admissible variations η.

Proof Let η be an admissible variation. Then $\hat{y}+\varepsilon\eta\in\mathscr{C}^1_\beta([a,b])$ for any $\varepsilon\in\mathbb{R}$, and

$$\begin{aligned}(\hat{y}+\varepsilon\eta)(a)&=\hat{y}(a)+\varepsilon\eta(a)\\&=\hat{y}(a)\\&=\alpha,\\(\hat{y}+\varepsilon\eta)(b)&=\hat{y}(b)+\varepsilon\eta(b)\\&=\hat{y}(b)\\&=\beta.\end{aligned}$$

Then, there exists $\delta>0$ such that

$$\mathscr{L}(y)\geq\mathscr{L}(\hat{y}) \quad for \quad all \quad y\in\mathscr{C}^1_{rd}([a,b])$$

with $y(a)=\alpha$, $y(b)=\beta$ and $\|y-\hat{y}\|<\delta$. Let $\varepsilon\in\mathbb{R}$ be such that

$$|\varepsilon|<\frac{\delta}{\|\eta\|}.$$

Hence,

$$\begin{aligned}\Phi(\varepsilon)&=\mathscr{L}(\hat{y}+\varepsilon\eta)\\&\geq\mathscr{L}(\hat{y})\\&=\Phi(0).\end{aligned}$$

Therefore, Φ has minimum at $\varepsilon=0$ and then

$$\Phi'(0)=0, \quad \Phi''(0)\geq 0.$$

This completes the proof. □

Theorem 5.2 (Sufficient Condition)
Let $\hat{y} \in \mathscr{C}^1_\beta([a,b])$ with $\hat{y}(a) = \alpha$, $\hat{y}(b) = \beta$. If

$$\mathscr{L}_1(\hat{y}, \eta) = 0 \quad \textit{and} \quad \mathscr{L}_2(\hat{y}, \eta) > 0$$

for all nontrivial admissible variations η, then $\hat{y}$ is a proper weak local minimum of the problem (5.9).

Proof Since

$$\mathscr{L}_2(\hat{y}, \eta) > 0,$$

there exists an $\varepsilon > 0$ with

$$\mathscr{L}_2(\hat{y}, \eta) \geq 2\varepsilon(b-a)\|\eta\|^2.$$

Because L is a C^2-function in x and v, there exists a $\delta > 0$ such that

$$|L''(t,z) - L''(t,\tilde{z})| < \varepsilon,$$

whenever $z, \tilde{z} \in \mathbb{R}^{2n}$ with $|z - \tilde{z}| < \delta$. Let $y \in \mathscr{C}^1([a,b])$ be such that $y \neq \hat{y}$, $y(a) = \alpha$, $y(b) = \beta$, and $\|y - \hat{y}\| < \delta$. We put

$$\eta = y - \hat{y}.$$

Note that η is a nontrivial admissible variation. By the Taylor formula, we get that there exists a $\zeta \in (0,1)$ such that for

$$z = \begin{pmatrix} \hat{y}^\beta + \zeta\eta^\beta \\ D_\beta\hat{y} + \zeta D_\beta\eta \end{pmatrix}^T, \quad \tilde{z} = \begin{pmatrix} \hat{y}^\beta \\ D_\beta\hat{y} \end{pmatrix},$$

and

$$\begin{aligned} |z(t) - \tilde{z}(t)| &\leq \zeta\|\eta\| \\ &< \|\eta\| \\ &< \delta \end{aligned}$$

holds for any $t \in [a,b]$, and

$$\begin{aligned} \mathscr{L}(y) - \mathscr{L}(\hat{y}) &= \Phi(1) - \Phi(0) \\ &= \Phi'(0) + \frac{1}{2}\Phi''(\zeta) \\ &= \frac{1}{2}\Phi''(\zeta) \end{aligned}$$

$$
\begin{aligned}
&= \frac{1}{2}\Phi''(0) + \frac{1}{2}\left(\Phi''(\zeta) - \Phi''(0)\right) \\
&\geq \frac{1}{2}\Phi''(0) - \frac{1}{2}|\Phi''(\zeta) - \Phi''(0)| \\
&= \frac{1}{2}\Phi''(0) - \frac{1}{2}\left|\int_a^b \begin{pmatrix} \eta^\beta(t) \\ D_\beta \eta(t) \end{pmatrix}^T (L''(t,z(t)) - L''(t,\tilde{z}(t))) \begin{pmatrix} \eta^\beta(t) \\ D_\beta \eta(t) \end{pmatrix} D_\beta t\right| \\
&\geq \varepsilon(b-a)\|\eta\|^2 - \frac{1}{2}\int_a^b \|\eta\|\varepsilon\|\eta\| D_\beta t \\
&= \varepsilon(b-a)\|\eta\|^2 - \frac{1}{2}\varepsilon(b-a)\|\eta\|^2 \\
&= \varepsilon\frac{b-a}{2}\|\eta\|^2 \\
&> 0.
\end{aligned}
$$

Therefore, $\hat{y}$ is a proper local minimum of (5.9). This completes the proof. □

By Theorem 5.1 and Theorem 5.2, we get

$$
\begin{aligned}
\mathscr{L}_1(y,\eta) = \int_a^b \Big(& L_x(t, y^\beta(t), D_\beta y(t))\eta^\beta(t) \\
& + L_v(t, y^\beta(t), D_\beta y(t)) \Big) D_\beta t,
\end{aligned}
\tag{5.10}
$$

and

$$
\begin{aligned}
\mathscr{L}_2(y,\eta) = \int_a^b \Big(& \left(\eta^\beta\right)^T(t) L_{xx}(t, y^\beta(t), D_\beta y(t))\eta^\beta(t) \\
& + 2\left(\eta^\beta\right)^T(t) L_{xv}(t, y^\beta(t), D_\beta y(t)) D_\beta \eta(t) \\
& + \left(\eta^{D_\beta}\right)^T(t) L_{vv}(t, y^\beta(t), D_\beta y(t)) D_\beta \eta(t) \Big) D_\beta t.
\end{aligned}
\tag{5.11}
$$

Example 5.1 Let $n = 1, I = 2^{\mathbb{N}_0}$. Consider the variational problem

$$\mathscr{L}(y) = \int_1^8 \left(t^2 + \left(y^\beta(t)\right)^2 + (D_\beta y(t))^3\right) d_\beta t \longrightarrow \min, \quad y(1) = y(8) = 0.$$

Here

$$L(t,x,v) = t^2 + x^2 + v^3.$$

Then

$$L_x(t,x,v) = 2x,$$

$$L_v(t,x,v) = 3v^2,$$

$$L_{xx}(t,x,v) = 2,$$

$$L_{xv}(t,x,v) = 0,$$

$$L_{vv}(t,x,v) = 6v.$$

Equations (5.10) and (5.11) take the form

$$\mathscr{L}_1(y,\eta) = \int_1^8 \left(2y^\beta(t)\eta^\beta(t) + 3\left(D_\beta y(t)\right)^2 \eta^{D_\beta}(t)\right) d_\beta t,$$

$$\mathscr{L}_2(y,\eta) = \int_1^8 \left(2\left(\eta^\beta(t)\right)^2 + 6D_\beta y(t)\left(D_\beta \eta(t)\right)^2\right) D_\beta t,$$

respectively.

Example 5.2 Let $n = 2, I = 2\mathbb{Z}$. Consider the variational problem

$$\mathscr{L}(y) = \int_0^{10} \left(ty_1^\beta(t) + \left(y_2^\beta(t)\right)^2 + (D_\beta y_1(t))\left(D_\beta y_2(t)\right)^2\right) D_\beta t \longrightarrow \quad \min,$$

$$y_1(0) = 1, \quad y_2(0) = 3,$$

$$y_1(0) = 4, \quad y_2(0) = 5.$$

Here

$$L(t,x_1,x_2,v_1,v_2) = tx_1 + x_2^2 + v_1 v_2^2.$$

We have

$$L_{x_1}(t,x_1,x_2,v_1,v_2) = t,$$

$$L_{x_2}(t,x_1,x_2,v_1,v_2) = 2x_2,$$

$$L_{v_1}(t,x_1,x_2,v_1,v_2) = v_2^2,$$

$$L_{v_)}2(t,x_1,x_2,v_1,v_2) = 2v_1v_2,$$

$$L_{x_1x_1}(t,x_1,x_2,v_1,v_2) = 0,$$

$$L_{x_1x_2}(t,x_1,x_2,v_1,v_2) = 0,$$

$$L_{x_2x_2}(t,x_1,x_2,v_1,v_2) = 2,$$

$$L_{x_1v_1}(t,x_1,x_2,v_1,v_2) = 0,$$

$$L_{x_1v_2}(t,x_1,x_2,v_1,v_2) = 0,$$

$$L_{x_2v_1}(t,x_1,x_2,v_1,v_2) = 0,$$

$$L_{x_2v_2}(t,x_1,x_2,v_1,v_2) = 0,$$

$$L_{v_1v_1}(t,x_1,x_2,v_1,v_2) = 0,$$

$$L_{v_1v_2}(t,x_1,x_2,v_1,v_2) = 2v_2,$$

$$L_{v_2v_2}(t,x_1,x_2,v_1,v_2) = 2v_1.$$

Then

$$L_x\left(t,y^\beta(t),D_\beta y(t)\right)\eta^\beta(t)+L_v\left(t,y^\beta(t),D_\beta y(t)\right)D_\beta\eta(t)$$

$$= \left(t,2y_2^\beta(t)\right)\begin{pmatrix}\eta_1^\beta(t)\\ \eta_2^\beta(t)\end{pmatrix}+\left(\left(D_\beta y_2(t)\right)^2,2D_\beta y_1(t)D_\beta y_2(t)\right)\begin{pmatrix}D_\beta\eta_1(t)\\ D_\beta\eta_2(t)\end{pmatrix}$$

$$= t\eta_1^\beta(t)+2y_2^\beta(t)\eta_2^\beta(t)+\left(D_\beta y_2(t)\right)^2D_\beta\eta_1(t)+2D_\beta y_1(t)D_\beta y_2(t)D_\beta\eta_2(t)$$

and equation (5.10) takes the form

$$\mathscr{L}_1(y,\eta) = \int_0^{10} \left(t\eta_1^\beta(t) + 2y_2^\beta(t)\eta_2^\beta(t) + (D_\beta y_2)^2 D_\beta \eta_1(t) \right.$$

$$\left. +2D_\beta y_1(t) D_\beta y_2(t) D_\beta \eta_2(t) \right) D_\beta t.$$

Next,

$$L_{xx}\left(t, y^\beta(t), D_\beta y(t)\right) = \begin{pmatrix} 0 & 0 \\ 0 & 2 \end{pmatrix},$$

and hence,

$$\left(\eta^\beta(t)\right)^T L_{xx}\left(t, y^\beta(t), D_\beta y(t)\right) \eta^\beta(t) = \left(\eta_1^\beta(t), \eta_2^\beta(t)\right) \begin{pmatrix} 0 & 0 \\ 0 & 2 \end{pmatrix} \begin{pmatrix} \eta_1^\beta(t) \\ \eta_2^\beta(t) \end{pmatrix}$$

$$= \left(\eta_1^\beta(t), \eta_2^\beta(t)\right) \begin{pmatrix} 0 \\ 2\eta_2^\beta(t) \end{pmatrix}$$

$$= 2\left(\eta_2^\beta(t)\right)^2,$$

and

$$L_{xv}\left(t, y^\beta(t), D_\beta y(t)\right) = \begin{pmatrix} 0 & 0 \\ 0 & 0 \end{pmatrix},$$

$$2\left(\eta^\beta(t)\right)^T L_{xv}\left(t, y^\beta(t), D_\beta y(t)\right) D_\beta \eta(t) = 0,$$

$$L_{vv}\left(t, y^\beta(t), D_\beta yt)\right) = \begin{pmatrix} 0 & 2D_\beta y_2(t) \\ 2D_\beta y_2(t) & 2D_\beta y_1(t) \end{pmatrix},$$

$$(D_\beta \eta(t))^T L_{vv}\left(t, y^\beta(t), D_\beta y(t)\right) D_\beta \eta(t)$$

$$= (D_\beta \eta_1(t), D_\beta \eta_2(t)) \begin{pmatrix} 0 & 2D_\beta y_2(t) \\ 2D_\beta y_2(t) & 2D_\beta y_1(t) \end{pmatrix} \begin{pmatrix} D_\beta \eta_1(t) \\ D_\beta \eta_2(t) \end{pmatrix}$$

$$= (D_\beta \eta_1(t), D_\beta \eta_2(t)) \begin{pmatrix} 2D_\beta y_2(t) D_\beta \eta_2(t) \\ 2D_\beta y_2(t) D_\beta \eta_1(t) + 2D_\beta y_1(t) D_\beta \eta_2(t) \end{pmatrix}$$

$$= 2D_\beta y_2(t) D_\beta \eta_1(t) D_\beta \eta_2(t) + 2D_\beta y_2(t) D_\beta \eta_1(t) D_\beta \eta_2(t)$$

$$+2D_\beta y_1(t) \left(D_\beta \eta_2(t)\right)^2$$

$$= 4D_\beta y_2(t) D_\beta \eta_1(t) D_\beta \eta_2(t) + 2D_\beta y_1(t) \left(D_\beta \eta_2(t)\right)^2.$$

Equation (5.11) takes the form

$$\mathscr{L}_2(y,\eta) = \int_0^{10} \left(2\left(\eta_2^\beta(t)\right)^2 + 4D_\beta y_2(t) D_\beta \eta_1(t) D_\beta \eta_2(t) + 2D_\beta y_1(t)\left(D_\beta \eta_2(t)\right)^2\right) D_\beta t.$$

Example 5.3 Let $n = 1$, $I = \mathbb{N}_0$. Consider the variational problem

$$\mathscr{L}(y) = \int_{-2}^{2} \left(t y^\beta(t) D_\beta y(t) + \left(D_\beta y(t)\right)^2\right) D_\beta t \longrightarrow \min,$$

$$y(2) = 1, \quad y(-2) = 4.$$

We have

$$L(t,x,v) = txv + v^2,$$

$$L_x(t,x,v) = tv,$$

$$L_v(t,x,v) = tx + 2v,$$

$$L_{xx}(t,x,v) = 0,$$

$$L_{xv}(t,x,v) = t,$$

$$L_{vv}(t,x,v) = 2.$$

Equations (5.10) and (5.11) take the form

$$\mathscr{L}_1(y,\eta) = \int_{-2}^{2} \left(t D_\beta y(t) \eta^\beta(t) + \left(t y^\beta(t) 2 D_\beta y(t)\right) D_\beta \eta(t)\right) D_\beta t,$$

$$\mathscr{L}_2(y,\eta) = \int_{-2}^{2} \left(2t\eta^\beta(t) D_\beta \eta(t) + 2D_\beta \eta(t)\right) D_\beta t$$

$$= 2\int_{-2}^{2} t\eta^\beta(t) D_\beta \eta(t) D_\beta t + 2\eta(t)\Big|_{t=-2}^{t=2}$$

$$= 2\int_{-2}^{2} t\eta^\beta(t) D_\beta \eta(t) D_\beta t,$$

respectively.

Exercise 5.1 Let $n = 1, I = \mathbb{N}_0^2$. Write equations (5.10) and (5.11) for the variational problem

$$\mathscr{L}(y) = \int_0^{49} t y^\beta(t) D_\beta y(t) D_\beta t \longrightarrow \min, \quad y(0) = y(49) = 1.$$

5.3 The Euler-Lagrange Equation

Theorem 5.3 *If $\hat{y}$ is a local minimum of* (5.9), *then the Euler-Lagrange equation*

$$D_\beta L_v(t, \hat{y}^\beta(t), D_\beta \hat{y}(t)) = L_x(t, \hat{y}^\beta(t), D_\beta \hat{y}(t)), \quad t \in [a, b], \tag{5.12}$$

holds, where

$$D_\beta L_v^{D_\beta} = \frac{D_\beta}{D_\beta t}(L_v)$$

Proof Let $\hat{y}$ be a local minimum of (5.9) and η be an admissible variation. Then

$$\begin{aligned}
0 &= \Phi(0) \\
&= \mathscr{L}_1(\hat{y}, \eta) \\
&= \int_a^b L_x(t, \hat{y}^\beta(t), D_\beta \hat{y}(t)) \eta^\beta(t) D_\beta t \\
&\quad + \int_a^b L_v(t, \hat{y}^\beta(t), D_\beta \hat{y}(t)) D_\beta \eta(t) D_\beta t \\
&= \left(\int_a^t L_x\left(s, \hat{y}^\beta(s), D_\beta \hat{y}(s)\right) D_\beta s \right) \eta(t) \Big|_{t=a}^{t=b} \\
&\quad - \int_a^b \left(\int_a^t L_x\left(s, \hat{y}^\beta(s), D_\beta \hat{y}(s)\right) D_\beta \eta(t) \right) D_\beta t \\
&\quad + \int_a^b L_v\left(t, \hat{y}^\beta(t), D_\beta \hat{y}(t)\right) D_\beta \eta(t) D_\beta t \\
&= - \int_a^b \left(\int_a^t L_x\left(s, \hat{y}^\beta(s), D_\beta \hat{y}(s)\right) D_\beta \eta(t) \right) D_\beta t \\
&\quad + \int_a^b L_v\left(t, \hat{y}^\beta(t), D_\beta \hat{y}(t)\right) D_\beta \eta(t) D_\beta t.
\end{aligned}$$

From this and, we get

$$L_x(t,\hat{y}^\beta(t),D_\beta\hat{y}(t)) = D_\beta L_v(t,\hat{y}^\beta(t),D_\beta\hat{y}(t)), \quad t\in[a,b].$$

This completes the proof. □

Definition 5.5 Equations (5.12) are called the Euler-Lagrange equations.

Example 5.4 Let $n=1$. We will find the solution of the problem

$$\int_a^b \sqrt{1+(D_\beta y(t))^2}D_\beta t \longrightarrow \min, \quad y(a)=0, \quad y(b)=1.$$

Here

$$L(t,x,v)=\sqrt{1+v^2}.$$

Then

$$L_x(t,x,v)=0,$$

$$L_v(t,x,v)=\frac{v}{\sqrt{1+v^2}}.$$

Therefore,

$$D_\beta\left(\frac{D_\beta y(t)}{\sqrt{1+(D_\beta y(t))^2}}\right)=0, \quad t\in[a,b],$$

whereupon

$$\frac{D_\beta y(t)}{\sqrt{1+(D_\beta y(t))^2}}=c, \quad t\in[a,b],$$

and

$$D_\beta y(t)=c_1, \quad y(t)=c_1t+c_2, \quad t\in[a,b].$$

Here c, c_1 and c_2 are constants. Note that

$$y(a) = c_1a+c_2=0,$$

$$y(b) = c_1b+c_2=1.$$

Then

$$c_1=\frac{1}{b-a}, \quad c_2=-\frac{a}{b-a}.$$

Consequently

$$y(t)=\frac{t-a}{b-a}, \quad t\in[a,b].$$

Example 5.5 Let $n = 1$ and $I = 2^{\mathbb{N}_0} \bigcup \{0\}, \beta(t) = 2t, t \in I$. We will find the solution of the variational problem

$$\int_1^8 \left(t^2 + y^\beta(t)\right) D_\beta y(t) D_\beta t \longrightarrow \min, \quad y(1) = y(8) = 1.$$

Here

$$\beta(t) = 2t,$$

$$\beta(0) = 0,$$

$$L(t,x,v) = (t^2 + x)v.$$

Then

$$L_x(t,x,v) = v,$$

$$L_v(t,x,v) = x + t^2.$$

Using the Euler-Lagrange equation, we get

$$D_\beta \left(t^2 + y^\beta(t)\right) = D_\beta y(t), \quad t \in [1,8],$$

or

$$D_\beta \left(t^2 + y(t) + tD_\beta y(t)\right) = D_\beta y(t), \quad t \in [1,8],$$

or

$$\beta(t) + t + D_\beta y(t) + D_\beta \left(tD_\beta y(t)\right) = D_\beta y(t), \quad t \in [1,8],$$

or

$$D_\beta \left(tD_\beta y(t)\right) = -3t, \quad t \in [1,8].$$

Hence,

$$tD_\beta y(t) = -t^2 + c_1, \quad t \in [1,8],$$

or

$$D_\beta y(t) = -t + \frac{c_1}{t}, \quad t \in [1,8].$$

Therefore,

$$y(t) - y(1) = -\int_1^t \tau D_\beta \tau + c_1 \int_1^t \frac{D_\beta \tau}{\tau}, \quad t \in [1,8],$$

or

$$y(t) = 1 - \frac{1}{3}\tau^2\Big|_{\tau=1}^{\tau=t} + c_1 \int_1^t \frac{D_\beta \tau}{\tau}, \quad t \in [1,8],$$

or

$$y(t) = 1 - \frac{1}{3}t^2 + \frac{1}{3} + c_1 \int_1^t \frac{D_\beta \tau}{\tau}, \quad t \in [1,8],$$

or

$$y(t) = \frac{4}{3} - \frac{1}{3}t^2 + c_1 \int_1^t \frac{D_\beta \tau}{\tau}, \quad t \in [1,8].$$

Here c_1 is a constant which will be determined by the condition $y(8) = 1$. We have

$$\frac{4}{3} - \frac{64}{3} + c_1 \int_1^8 \frac{D_\beta \tau}{\tau} = 1,$$

or

$$c_1 \int_1^8 \frac{D_\beta \tau}{\tau} = 21,$$

or

$$c_1 = \frac{21}{\int_1^8 \frac{D_\beta \tau}{\tau}}.$$

Consequently,

$$y(t) = \frac{4-t^3}{3} + \frac{21 \int_1^t \frac{D_\beta \tau}{\tau}}{\int_1^8 \frac{D_\beta \tau}{\tau}}, \quad t \in [1,8].$$

Example 5.6 Let $I = 3\mathbb{N}$, $\beta(t) = t+3$, $t \neq 0$, $\beta(0) = 0$. We will find the solution of the variational problem

$$\int_0^9 \left(-(D_\beta y_1(t))^2 + y_2^\beta(t) D_\beta y_1(t) + t y_1^\beta(t) + (D_\beta y_2(t))^2 \right) D_\beta t \longrightarrow \min,$$

$$\begin{pmatrix} y_1(0) \\ y_2(0) \end{pmatrix} = \begin{pmatrix} 9 \\ 1 \end{pmatrix}, \quad \begin{pmatrix} y_1(9) \\ y_2(9) \end{pmatrix} = \begin{pmatrix} -1 \\ 1 \end{pmatrix}.$$

Here

$$L(t, x_1, x_2, v_1, v_2) = -v_1^2 + x_2 v_1 + t x_1 + v_2^2.$$

Then

$$L_{x_1}(t,x_1,x_2,v_1,v_2) = t,$$

$$L_{x_2}(t,x_1,x_2,v_1,v_2) = v_1,$$

$$L_{v_1}(t,x_1,x_2,v_1,v_2) = x_2 - 2v_1,$$

$$L_{v_2}(t,x_1,x_2,v_1,v_2) = 2v_2.$$

Then the Euler-Lagrange equations take the form

$$\begin{cases} D_\beta\left(y_2^\beta(t) - 2D_\beta y_1(t)\right) = t \\ 2D_\beta\left(D_\beta y_2(t)\right) = D_\beta y_1(t), \quad t \in [0,9]. \end{cases}$$

Let

$$f(t) = \frac{1}{2}t^2 - \frac{3}{2}t, \quad t \in [0,9].$$

Then

$$\begin{aligned} D_\beta f(t) &= \frac{1}{2}(\beta(t)+t) - \frac{3}{2} \\ &= \frac{1}{2}(t+3+t) - \frac{3}{2} \\ &= t. \end{aligned}$$

Therefore,

$$\begin{cases} y_2^\beta(t) - 2D_\beta y_1(t) = \frac{1}{2}t^2 - \frac{3}{2}t + c_1 \\ 2D_\beta y_2(t) = y_1(t) + c_2, \quad t \in [0,9], \end{cases}$$

or

$$\begin{cases} y_2(t) + 3D_\beta y_2(t) - 4D_\beta^2 y_2(t) = \frac{1}{2}t^2 - \frac{3}{2}t + c_1 \\ y_1(t) = 2D_\beta y_2(t) - c_2, \quad t \in [0,9], \end{cases}$$

or

$$\begin{cases} 4D_\beta^2 y_2(t) - 3D_\beta y_2(t) - y_2(t) = -\frac{1}{2}t^2 + \frac{3}{2}t + c_1 \\ y_1(t) = 2D_\beta y_2(t) - c_2, \quad t \in [0,9]. \end{cases}$$

Here c_1 and c_2 are constants. Consider the equation

$$4D_\beta^2 y_2(t) - 3D_\beta y_2(t) - y_2(t) = -\frac{1}{2}t^2 + \frac{3}{2}t + c_1, \quad t \in [0,9]. \tag{5.13}$$

Note that

$$y_{2h}(t) = a_1 e_{-\frac{1}{4},\beta}(t,0) + a_2 e_{1,\beta}(t,0), \quad t \in [0,9],$$

is a solution of the corresponding homogeneous equation of equation (5.13). Here a_1 and a_2 are constants. We will search a solution of equation (5.13) in the form

$$y_2(t) = a_1(t) e_{-\frac{1}{4},\beta}(t,0) + a_2(t) e_{1,\beta}(t,0), \quad t \in [0,9],$$

where $a_1, a_2 \in \mathscr{C}^1([0,9])$ will be determined below. We have

$$\begin{aligned} D_\beta y_2(t) &= D_\beta a_1(t) e_{-\frac{1}{4},\beta}(t+3,0) + D_\beta a_2(t) e_{1,\beta}(t+3,0) \\ &\quad -\frac{1}{4} a_1(t) e_{-\frac{1}{4},\beta}(t,0) + a_2(t) e_{1,\beta}(t,0), \quad t \in [0,9]. \end{aligned}$$

We want

$$D_\beta a_1(t) e_{-\frac{1}{4},\beta}(t+3,0) + D_\beta a_2(t) e_{1,\beta}(t+3,0) = 0, \quad t \in [0,9].$$

Then

$$D_\beta y_2(t) = -\frac{1}{4} a_1(t) e_{-\frac{1}{4},\beta}(t,0) + a_2(t) e_{1,\beta}(t,0), \quad t \in [0,9].$$

Hence,

$$\begin{aligned} D_\beta^2 y_2(t) &= -\frac{1}{4} D_\beta a_1(t) e_{-\frac{1}{4},\beta}(t+3,0) + D_\beta a_2(t) e_{1,\beta}(t+3,0) \\ &\quad + \frac{1}{16} a_1(t) e_{-\frac{1}{4},\beta}(t,0) + a_2(t) e_{1,\beta}(t,0), \quad t \in [0,9], \end{aligned}$$

and

$$\begin{aligned} 4D_\beta^2 y_2(t) - 3D_\beta y_2(t) - y_2(t) &= -D_\beta a_1(t) e_{-\frac{1}{4},\beta}(t+3,0) + 4D_\beta a_2(t) e_{1,\beta}(t+3,0) \\ &\quad + \frac{1}{4} a_1(t) e_{-\frac{1}{4},\beta}(t,0) + 4a_2(t) e_{1,\beta}(t,0) \\ &\quad + \frac{3}{4} a_1(t) e_{-\frac{1}{4},\beta}(t,0) - 3a_2(t) e_{1,\beta}(t,0) \\ &\quad - a_1(t) e_{-\frac{1}{4},\beta}(t,0) - a_2(t) e_{1,\beta}(t,0) \\ &= \frac{1}{2}t^2 - \frac{3}{2}t + c_1, \quad t \in [0,9], \end{aligned}$$

or

$$-D_\beta a_1(t)e_{-\frac{1}{4},\beta}(t+3,0)+4D_\beta a_2(t)e_{1,\beta}(t+3,0)=\frac{1}{2}t^2-\frac{3}{2}t+c_1,$$

$t\in[0,9]$. Thus we get the system

$$\begin{cases} D_\beta a_1(t)e_{-\frac{1}{4},\beta}(t+3,0)+D_\beta a_2(t)e_{1,\beta}(t+3,0)=0 \\ -D_\beta a_1(t)e_{-\frac{1}{4},\beta}(t+3,0)+4D_\beta a_2(t)e_{1,\beta}(t+3,0)=\frac{1}{2}t^2-\frac{3}{2}t+c_1, \quad t\in[0,9]. \end{cases}$$

Consequently,

$$D_\beta a_2(t)=\frac{1}{5}\left(\frac{1}{2}t^2-\frac{3}{2}t+c_1\right)e_{1,\beta}(0,t+3), \quad t\in[0,9],$$

and we take

$$a_2(t)=\frac{1}{5}\int_0^9\left(\frac{1}{2}\tau^2-\frac{3}{2}\tau+c_1\right)e_{1,\beta}(0,\tau+3)D_\beta\tau, \quad t\in[0,9],$$

and

$$\begin{aligned} D_\beta a_1(t) &= -D_\beta a_2(t)e_{1,\beta}(t+3,0)e_{-\frac{1}{4},\beta}(0,t+3) \\ &= -\frac{1}{5}\left(\frac{1}{2}t^2-\frac{3}{2}t+c_1\right)e_{1,\beta}(0,t+3)e_{1,\beta}(t+3,0)e_{-\frac{1}{4},\beta}(0,t+3) \\ &= -\frac{1}{5}\left(\frac{1}{2}t^2-\frac{3}{2}t+c_1\right)e_{-\frac{1}{4},\beta}(0,t+3), \quad t\in[0,9]. \end{aligned}$$

We take

$$a_1(t)=-\frac{1}{5}\int_0^t\left(\frac{1}{2}\tau^2-\frac{3}{2}\tau+c_1\right)e_{-\frac{1}{4},\beta}(0,\tau+3)D_\beta\tau, \quad t\in[0,9].$$

Therefore,

$$\begin{aligned} & a_1(t)e_{-\frac{1}{4},\beta}(t,0)+a_2(t)e_{1,\beta}(t,0) \\ &= -\frac{1}{5}\int_0^t\left(\frac{1}{2}\tau^2-\frac{3}{2}\tau+c_1\right)e_{-\frac{1}{4},\beta}(t,\tau+3)D_\beta\tau \\ &\quad +\frac{1}{5}\int_0^t\left(\frac{1}{2}\tau^2-\frac{3}{2}\tau+c_1\right)e_{1,\beta}(t,\tau+3)D_\beta\tau \\ &= \frac{1}{5}\int_0^t\left(\frac{1}{2}\tau^2-\frac{3}{2}\tau+c_1\right)\left(e_{1,\beta}(t,\tau+3)-e_{-\frac{1}{4},\beta}(t+3,\tau+3)\right)D_\beta\tau, \quad t\in[0,9], \end{aligned}$$

and

$$y_2(t) = a_1 e_{-\frac{1}{4},\beta}(t,0) + a_2 e_{1,\beta}(t,0)$$

$$+\frac{1}{5}\int_0^t \left(\frac{1}{2}\tau^2 - \frac{3}{2}\tau + c_1\right)\left(e_{1,\beta}(t,\tau+3) - e_{-\frac{1}{4},\beta}(t,\tau+3)\right) D_\beta \tau,$$

$$D_\beta y_2(t) = -\frac{1}{4}a_1 e_{-\frac{1}{4},\beta}(t,0) + a_2 e_{1,\beta}(t,0)$$

$$+\frac{1}{5}\left(\frac{1}{2}t^2 - \frac{3}{2}t + c_1\right)\left(e_{1,\beta}(t+3,t+3) - e_{-\frac{1}{4},\beta}(t+3,t+3)\right)$$

$$+\frac{1}{5}\int_0^t \left(\frac{1}{2}\tau^2 - \frac{3}{2}\tau + c_1\right)\left(e_{1,\beta}(t,\tau+3) + \frac{1}{4}e_{-\frac{1}{4},\beta}(t,\tau+3)\right) D_\beta \tau,$$

$$y_1(t) = -\frac{1}{2}a_1 e_{-\frac{1}{4},\beta}(t,0) + 2a_2 e_{1,\beta}(t,0)$$

$$+\frac{2}{5}\int_0^t \left(\frac{1}{2}\tau^2 - \frac{3}{2}\tau + c_1\right)\left(e_{1,\beta}(t,\tau+3) + \frac{1}{4}e_{-\frac{1}{4},\beta}(t,\tau+3)\right) D_\beta \tau - c_2,$$

$t \in [0,9]$. For the constants a_1, a_2, c_1 and c_2 we have the system

$$\begin{cases} -\frac{1}{2}a_1 + 2a_2 - c_2 = 0 \\ a_1 + a_2 = 1 \\ -\frac{1}{2}a_1 e_{-\frac{1}{4},\beta}(9,0) + 2a_2 e_{1,\beta}(9,0) \\ +\frac{2}{5}\int_0^9 \left(\frac{1}{2}\tau^2 - \frac{3}{2}\tau + c_1\right)\left(e_{1,\beta}(9,\tau+3) + \frac{1}{4}e_{-\frac{1}{4},\beta}(9,\tau+3)\right) D_\beta \tau - c_2 = -1 \\ a_1 e_{-\frac{1}{4},\beta}(9,0) + a_2 e_{1,\beta}(9,0) \\ +\frac{1}{5}\int_0^2 \left(\frac{1}{2}\tau^2 - \frac{3}{2}\tau + c_1\right)\left(e_{1,\beta}(9,\tau+3) - e_{-\frac{1}{4},\beta}(9,\tau+3)\right) D_\beta \tau = 1. \end{cases}$$

Exercise 5.2 Let $I = 3^{\mathbb{N}_0}\beta(t) = 3t, t \neq 1, \beta(1) = 1$. Find the solution of the variational problem.

$$\int_1^{27} \left(\sqrt{1 + (D_\beta y_1(t))^2} + \sqrt{1 + (D_\beta y_2(t))^2} \right) D_\beta t \longrightarrow \min,$$

$$\begin{pmatrix} y_1(1) \\ y_2(1) \end{pmatrix} = \begin{pmatrix} 0 \\ 0 \end{pmatrix}, \quad \begin{pmatrix} y_1(27) \\ y_2(27) \end{pmatrix} = \begin{pmatrix} 1 \\ 1 \end{pmatrix}.$$

5.4 The Legendre Condition

We introduce the following notations:

$$P = L_{vv}(\cdot, \hat{y}^\beta, D_\beta \hat{y}), \quad Q = L_{xx}(\cdot, \hat{y}^\beta, D_\beta \hat{y}),$$

$$R = L_{xv}(\cdot, \hat{y}^\beta, D_\beta \hat{y}),$$

$$\alpha^\dagger = \frac{1}{\alpha} \quad if \quad \alpha \in \mathbb{R} \backslash \{0\}, \quad 0^\dagger = 0.$$

Theorem 5.4 (The Legendre Condition)
If $\hat{y}$ is a weak local solution of the variational problem (5.9), *then*

$$P(t) + (\beta(t) - t) \left(R(t) + R^T(t) + (\beta(t) - t)Q(t) + ((\beta - I)(\beta(t)))^\dagger P(\beta(t)) \right) \geq 0,$$

$t \in [a, b]^{\kappa^2}$.

Proof Let $s \in [a, b]^{\kappa^2}$.

1. Suppose that

$$s < \beta(s) < t.$$

We take $\gamma \in \mathbb{R}^n$ arbitrarily and define $\eta : [a, b] \to \mathbb{R}^n$ as follows.

$$\eta(t) = \begin{cases} \gamma\sqrt{(\beta - I)(s)} & if \quad t = \beta(s) \\ 0 & otherwise. \end{cases}$$

We have $\eta(a) = \eta(b) = 0$. Therefore, η is an admissible variation. Note that

$$\begin{aligned} D_\beta \eta(s) &= \frac{\eta(\beta(s)) - \eta(s)}{(\beta - I)(s)} \\ &= \frac{\gamma\sqrt{(\beta - I)(s)}}{(\beta - I)(s)} \\ &= \frac{\gamma}{\sqrt{(\beta - I)(s)}} \end{aligned}$$

and

$$\begin{aligned} D_\beta \eta(\beta(s)) &= \frac{\eta(\beta(\beta(s))) - \eta(\beta(s))}{(\beta - I)(\beta(s))} \\ &= -\frac{\gamma\sqrt{(\beta - I)(s)}}{(\beta - I)(\beta(s))}. \end{aligned}$$

Next,

$$D_\beta \eta(t) = 0 \quad for \quad all \quad t \in [a,b] \backslash \{s, \beta(s)\}.$$

Thus

$$\begin{aligned} &\int_a^{\beta(s)} \left(\left(\eta^\beta\right)^T Q\eta^\beta + 2\left(\eta^\beta\right)^T RD_\beta\eta + (D_\beta\eta)^T PD_\beta\eta \right)(t) D_\beta t \\ &= (\beta - I)(s) \left(\left(\eta^\beta\right)^T Q\eta^\beta + 2\left(\eta^\beta\right)^T RD_\beta\eta + (D_\beta\eta)^T PD_\beta\eta \right)(s) \\ &= (\beta - I)(s) \left(\gamma^T\sqrt{(\beta - I)} Q\gamma\sqrt{(\beta - I)} + 2\gamma^T\sqrt{(\beta - I)} R\frac{\gamma}{\sqrt{(\beta - I)}} \right. \\ &\qquad \left. + \frac{\gamma^T}{\sqrt{(\beta - I)}} P \frac{\gamma}{\sqrt{(\beta - I)}} \right)(s) \\ &= (\beta - I)(s) \left((\beta - I)\gamma^T Q\gamma + 2\gamma^T R\gamma + \frac{1}{(\beta - I)}\gamma^T P\gamma \right)(s) \\ &= \left(\gamma^T P\gamma + (\beta - I)^2\gamma^T Q\gamma + \gamma^T(\beta - I)R\gamma + \gamma(\beta - I)R^T\gamma\right)(s) \\ &= \gamma^T \left(P(s) + (\beta - I)(s)\left(R(s) + R^T(s) + (\beta - I)(s)Q(s)\right)\right)\gamma. \end{aligned}$$

a. Let $\beta(s)$ be right-dense. Then

$$\int_{\beta(s)}^{b} \left(\left(\eta^{\beta}\right)^{T} Q\eta^{\beta} + 2\left(\eta^{\beta}\right)^{T} RD_{\beta}\eta + (D_{\beta}\eta)^{T} PD_{\beta}\eta \right)(t)D_{\beta}t = 0.$$

b. Let $\beta(s)$ be right-scattered. Then

$$\begin{aligned}
&\int_{\beta(s)}^{b} \left(\left(\eta^{\beta}\right)^{T} Q\eta^{\beta} + 2\left(\eta^{\beta}\right)^{T} RD_{\beta}\eta + (D_{\beta}\eta)^{T} PD_{\beta}\eta \right)(t)D_{\beta}t \\
&= \int_{\beta(s)}^{\beta(\beta(s))} \left(\left(\eta^{\beta}\right)^{T} Q\eta^{\beta} + 2\left(\eta^{\beta}\right)^{T} RD_{\beta}\eta + (D_{\beta}\eta)^{T} PD_{\beta}\eta \right)(t)D_{\beta}t \\
&= (\beta - I)(\beta(s)) \left(\left(\eta^{\beta}\right)^{T} Q\eta^{\beta} + 2\left(\eta^{\beta}\right)^{T} RD_{\beta}\eta + (D_{\beta}\eta)^{T} PD_{\beta}\eta \right)(\beta(s)) \\
&= (\beta - I)(\beta(s)) \frac{(\beta - I)(s)}{((\beta - I)(\beta(s)))^{2}} \gamma^{T} P(\beta(s))\gamma \\
&= \frac{(\beta - I)(s)}{(\beta - I)(\beta(s))} \gamma^{T} P(\beta(s))\gamma.
\end{aligned}$$

Consequently,

$$\begin{aligned}
&\int_{a}^{b} \left(\left(\eta^{\beta}\right)^{T} Q\eta^{\beta} + 2\left(\eta^{\beta}\right)^{T} RD_{\beta}\eta + (D_{\beta}\eta)^{T} PD_{\beta}\eta \right)(t)D_{\beta}t \\
&= \begin{cases} \gamma^{T}\left(P(s) + (\beta - I)(s)\left(R(s) + R^{T}(s) + (\beta - I)(s)Q(s)\right)\right)\gamma & if \quad \beta(\beta(s)) = \beta(s) \\ \gamma^{T}\left(P(s) + (\beta - I)(s)\left(R(s) + R^{T}(s) + (\beta - I)(s)Q(s)\right)\right)\gamma & \\ + \dfrac{(\beta - I)(s)}{(\beta - I)(\beta(s))} \gamma^{T} P(\beta(s))\gamma & if \quad \beta(\beta(s)) > \beta(s) \end{cases} \\
&= \gamma^{T}\left(P(s) + (\beta - I)(s)\left(R(s) + R^{T}(s) + (\beta - I)(s)Q(s) + ((\beta - I)(\beta(s)))^{\dagger}\right)\right)\gamma.
\end{aligned}$$

From this and, we conclude that

$$\gamma^{T}\left(P(s) + (\beta - I)(s)\left(R(s) + R^{T}(s) + (\beta - I)(s)Q(s) + ((\beta - I)(\beta(s)))^{\dagger}\right)\right)\gamma \geq 0.$$

2. Let s be right-dense.

a. Let s be left-scattered. Therefore, there exists a decreasing sequence $\{s_k\}_{k\in\mathbb{N}} \subset [a,b]$ such that

$$\lim_{k\to\in fty} s_k = s.$$

For $\gamma \in \mathbb{R}^n$ and $k \in \mathbb{N}$, we define

$$\eta_k(t) = \begin{cases} \gamma \dfrac{s_k - t}{\sqrt{s_k - s}} & if \quad t \in [s, s_k] \\ 0 & otherwise. \end{cases}$$

Note that η_k is an admissible variation. We have

$$\int_a^s \left(\left(\eta^\beta\right)^T Q\eta^\beta + 2(\eta^\beta)^T RD_\beta \eta + (D_\beta \eta)^T PD_\beta \eta \right)(t) D_\beta t$$

$$= (\beta - I)(\beta^{-1}(s)) \left((\eta^\beta)^T Q\eta^\beta + 2(\eta^\beta)^T RD_\beta \eta + (D_\beta \eta)^T PD_\beta \eta \right)(\beta^{-1}(s))$$

$$= (\beta - I)(\beta^{-1}(s)) \Bigg((s_k - s)\gamma^T Q(\beta^{-1}(s))\gamma + 2\frac{s_k - s}{(\beta - I)(\beta^{-1}(s))} \gamma^T R(\beta^{-1}(s))\gamma$$

$$+ \frac{s_k - s}{((\beta - I)(\beta^{-1}(s)))^2} \gamma^T P(\beta^{-1}(s))\gamma \Bigg)$$

$$= (s_k - s)\gamma^T \left(2R(\beta^{-1}(s)) + (\beta - I)(\beta^{-1}(s)) Q(\beta^{-1}(s)) + \frac{1}{(\beta - I)(\beta^{-1}(s))} P(\beta^{-1}(s)) \right) \gamma$$

and

$$\int_s^b \left((\eta^\beta)^T Q\eta^\beta + 2(\eta^\beta)^T RD_\beta \eta + (D_\beta \eta)^T PD_\beta \eta \right)(t) D_\beta t$$

$$= \int_s^{s_k} \left(\frac{(s_k - \beta(t))^2}{s_k - s} \gamma^T Q(t)\gamma - 2\frac{s_k - \beta(t)}{s_k - s} \gamma^T R(t)\gamma + \frac{1}{s_k - s} \gamma^T P(t)\gamma \right) D_\beta t.$$

From this and, letting $k \to \in fty$, we get

$$\gamma^T P(s)\gamma \geq 0,$$

i.e.,

$$\gamma^T \left(P(s) + (\beta - I)(s) \left(R(s) + R^T(s) + (\beta - I)(s) Q(s) + ((\beta - I)(\beta(s)))^\dagger \right) \right) \gamma \geq 0.$$

b. Assume that s is left-dense. Then there exists an increasing sequence $\{z_k\}_{k \in \mathbb{N}} \subset [a, b]$ such that

$$\lim_{k \to \in fty} z_k = s.$$

For $\gamma \in \mathbb{R}^n$ and $k \in \mathbb{R}^n$ we define

$$\eta_k(t) = \begin{cases} \gamma \frac{s_k - t}{\sqrt{s_k - s}} & if \quad t \in [s, s_k] \\ \gamma \frac{t - z_k}{\sqrt{s - z_k}} & if \quad t \in [z_k, s] \\ 0 \quad otherwise. \end{cases}$$

Then η_k is an admissible variation. We have

$$\int_a^s \left(\left(\eta^\beta\right)^T Q\eta^\beta + 2(\eta^\beta)^T R D_\beta \eta + (D_\beta \eta)^T P D_\beta \eta \right)(t) D_\beta t$$

$$+ \int_s^b \left(\left(\eta^\beta\right)^T Q\eta^\beta + 2(\eta^\beta)^T R D_\beta \eta + (D_\beta \eta)^T P D_\beta \eta \right)(t) D_\beta t$$

$$= \int_{z_k}^s \left(\frac{(z_k - \beta(t))^2}{s - z_k} \gamma^T Q(t)\gamma - 2\frac{\beta(t) - z_k}{s - z_k} \gamma^T R(t)\gamma + \frac{1}{s - z_k} \gamma^T P(t)\gamma \right) D_\beta t$$

$$+ \int_s^{s_k} \left(\frac{(s_k - \beta(t))^2}{s_k - s} \gamma^T Q(t)\gamma - 2\frac{s_k - \beta(t)}{s_k - s} \gamma^T R(t)\gamma + \frac{1}{s_k - s} \gamma^T P(t)\gamma \right) D_\beta t.$$

Hence, using Theorem 5.1 and letting $k \to \in fty$, we obtain

$$\gamma^T P(s)\gamma \geq 0$$

and

$$\gamma^T \left(P(s) + (\beta - I)(s) \left(R(s) + R^T(s) + (\beta - I)(s)Q(s) + ((\beta - I)(\beta(s)))^\dagger \right) \right) \gamma \geq 0.$$

This completes the proof. □

5.5 The Jacobi Condition

Theorem 5.5 *Suppose that*

$$P(t) \quad and \quad P(t) + (\beta(t) - t)R(t) \quad are \quad invertible \quad for \quad all \quad t \in [a, b]. \tag{5.14}$$

Then

$$\mathscr{L}_2(\hat{y},\eta) = \int_a^b \left(\left(\eta^\beta\right)^T C\eta^\beta + \xi^T B\xi\right)(t)D_\beta t,$$

where

$$\xi = PD_\beta\eta + R^T\eta^\beta, \quad B = P^{-1}, \quad C = Q - RP^{-1}R^T.$$

Proof We have

$$\begin{aligned}
\mathscr{L}_2(\hat{y},\eta) &= \int_a^b \left(\left(\eta^\beta\right)^T Q\eta^\beta + 2\left(\eta^\beta\right)^T RD_\beta\eta + \left(D_\beta\eta\right)^T PD_\beta\eta\right)(t)D_\beta t \\
&= \int_a^b \Bigg(\left(\eta^\beta\right)^T \left(C + RBR^T\right)\eta^\beta + 2\left(\eta^\beta\right)^T RBPD_\beta\eta \\
&\quad + \left(PD_\beta\eta\right)^T B\left(PD_\beta\eta\right)\Bigg)(t)D_\beta t \\
&= \int_a^b \left(\left(\eta^\beta\right)^T C\eta^\beta + \left(PD_\beta\eta + R^T\eta^\beta\right)^T B\left(PD_\beta\eta + R^T\eta^\beta\right)\right)(t)D_\beta t \\
&= \int_a^b \left(\left(\eta^\beta\right)^T C\eta^\beta + \xi^T B\xi\right)(t)D_\beta t.
\end{aligned}$$

This completes the proof. □

We introduce the following notations:

$$A = -P^{-1}R^T, \quad B = P^{-1}, \quad C = Q - RP^{-1}R^T.$$

Then

$$\begin{aligned}
\xi &= PD_\beta\eta + R^T\eta^\beta \quad \Longleftrightarrow \\
B\xi &= P^{-1}\xi \\
&= D_\beta\eta + P^{-1}R^T\eta^\beta \\
&= D_\beta\eta - A\eta^\beta,
\end{aligned}$$

i.e.,

$$D_\beta\eta = A\eta^\beta + B\xi.$$

Theorem 5.6 (The Jacobi Condition)

Assume that (5.14) *hold. Then* $\mathscr{L}_2$ *is positive definite if and only if the linear Hamiltonian system*

$$\begin{cases} D_\beta \eta = A(t)\eta^\beta + B(t)\xi \\ D_\beta \xi = C(t)\eta^\beta - A^T(t)\xi \end{cases}$$

is disconjugate on $[a,b]$.

Proof We have

$$\begin{aligned} I-(\beta-I)A &= I+(\beta-I)P^{-1}R^T \\ &= P^{-1}\left(P+(\beta-I)R^T\right) \\ &= \left((P+(\beta-I)R)P^{-1}\right)^T \end{aligned}$$

is invertible. Then the assertion follows from the results in the previous section. This completes the proof. □

Theorem 5.7 *Let* R *be symmetric, differentiable and invertible, and* $P+(\beta-I)R$ *be invertible for all* $t\in[a,b]$. *Then* $\mathscr{L}_2$ *is positive definite if and only if the linear Hamiltonian system*

$$\begin{cases} D_\beta \eta = \tilde{P}^{-1}(t)\xi \\ D_\beta \xi = \tilde{Q}(t)\eta^\beta \end{cases}$$

is disconjugate on $[a,b]$, *where*

$$\tilde{Q} = Q - D_\beta R, \quad \tilde{P} = P + (\beta - I)R.$$

Proof We have

$$\begin{aligned} D_\beta\left(\eta^T R\eta\right) &= \left(\eta^\beta\right)^T D_\beta R\eta^\beta + \left(\eta^\beta\right)^T RD_\beta\eta + \left(D_\beta\eta\right)^T R\eta \\ &= \left(\eta^\beta\right)^T D_\beta R\eta^\beta + 2\left(\eta^\beta\right)^T RD_\beta\eta - (\beta-I)\left(D_\beta\eta\right)^T RD_\beta\eta. \end{aligned}$$

Hence,

$$\mathscr{L}_2(y,\eta) = \int_a^b \Bigg(\left(\left(\eta^\beta\right)^T Q\eta^\beta + D_\beta\left(\eta^T R\eta\right) - \left(\eta^\beta\right)^T D_\beta R\eta^\beta \right.$$

$$\left. +(\beta - I)\left(D_\beta\eta\right)^T RD_\beta\eta + \left(D_\beta\eta\right)^T PD_\beta\eta \right)(t)D_\beta t$$

$$= \int_a^b \left(\left(\eta^\beta\right)^T (Q - D_\beta R)\eta^\beta + \left(D_\beta\eta\right)^T ((\beta - I)R + P)D_\beta\eta \right)(t)D_\beta t$$

$$= \int_a^b \left(\left(\eta^\beta\right)^T \tilde{Q}\eta^\beta + \left(D_\beta\eta\right)^T \tilde{P}D_\beta\eta \right)(t)D_\beta t,$$

whereupon, using the results in the previous chapter, the assertion follows. This completes the proof. □

5.6 Advanced Practical Problems

Problem 5.1 Let $n = 1$, $I = 3^{\mathbb{N}_0}\beta(t) = 3t, t \neq 1, \beta(1) = 1$. Write equations (5.10) and (5.11) for the variational problem

$$\mathscr{L}(y) = \int_1^{27} \left(t^2 + \left(D_\beta y(t)\right)^2\right) D_\beta t \longrightarrow \min, \quad y(1) = y(27) = 3.$$

Answer.

$$\mathscr{L}_1(y,\eta) = 2\int_1^{27} D_\beta y(t) D_\beta \eta(t) D_\beta t,$$

$$\mathscr{L}_2(y,\eta) = 2\int_1^{27} \left(D_\beta\eta(t)\right)^2 D_\beta t.$$

Problem 5.2 Write the Euler-Lagrange equations for the following variational problems:

1. $\int_0^{10} \left(ty^\beta(t)D_\beta y(t) + \left(D_\beta y(t)\right)^3\right) D_\beta t \longrightarrow \min, \quad y(0) = 1, \quad y(10) = -2, \quad I = \mathbb{N}\beta(t) = t + 1, t \neq 1, \beta(1) = 1,$

2. $\int_{1}^{16} \left(t y_1^{\beta}(t) y_2^{\beta}(t) + D_{\beta} y_1(t) D_{\beta} y_2(t) \right) D_{\beta} t \longrightarrow \min,$

$$\begin{pmatrix} y_1(1) \\ y_2(1) \end{pmatrix} = \begin{pmatrix} 0 \\ -1 \end{pmatrix}, \quad \begin{pmatrix} y_1(16) \\ y_2(16) \end{pmatrix} = \begin{pmatrix} 1 \\ 1 \end{pmatrix}, \quad I = 2^{\mathbb{N}_0} \beta(t) = 2t, t \neq 1, \beta(1) = 1.$$

Problem 5.3 Check the Legendre condition for the following variational problems:

1. $\int_{1}^{9} \left(\left(y^{\beta}(t) \right)^2 - 3 \left(D_{\beta} y(t) \right)^2 \right) D_{\beta} t \longrightarrow \min,$

$$y(1) = y(9) = 1, \quad I = 3^{\mathbb{N}_0} \beta(t) = 3t, t \neq 1, \beta(1) = 1.$$

2. $\int_{0}^{20} \left(y^{\beta}(t) D_{\beta} y(t) + y^{\beta}(t) \left(D_{\beta} y(t) \right)^3 + t^2 \right) d_{\beta} t \longrightarrow \min,$

$$y(0) = 1, \quad y(20) = 3, \quad I = 2\mathbb{N} \beta(t) = t + 2, t \neq 1, \beta(1) = 1.$$

Problem 5.4 Check the Euler condition for the following variational problems:

1. $\int_{-1}^{10} \left(\left(t^2 y^{\beta}(t) \right)^3 - 3(t+1) \left(D_{\beta} y(t) \right)^4 \right) D_{\beta} t \longrightarrow \min,$

$$y(-1) = 0, \quad y(10) = 25, \quad I = \mathbb{N} \beta(t) = t + 1, t \neq 1, \beta(1) = 1.$$

2. $\int_{1}^{8} \left(y^{\beta}(t) + \left(D_{\beta} y \right)^3 (t) + y^{\beta}(t) \left(D_{\beta} y(t) \right)^7 + t^3 + t^2 - 2t + 1 \right) D_{\beta} t \longrightarrow \min,$

$$y(1) = 0, \quad y(8) = 4, \quad I = 2^{\mathbb{N}_0} \beta(t) = 2t, t \neq 1, \beta(1) = 1.$$

Problem 5.5 Check the conditions of Theorem 5.7 for the following variational problems:

1. $\int_{1}^{27} \left(\left(t^2 - 7 y^{\beta}(t) \right)^3 + t \left(D_{\beta} y(t) \right)^2 - t \right) D_{\beta} t \longrightarrow \min,$

$$y(1) = 0, \quad y(27) = 2, \quad I = 3^{\mathbb{N}_0} \beta(t) = 3t, t \neq 1, \beta(1) = 1.$$

2. $\int_{1}^{8} \left(y^{\beta}(t) + D_{\beta} y(t) + t y^{\beta}(t) \left(D_{\beta} y(t) \right)^2 + t \right) d_{\beta} t \longrightarrow \min,$

$$y(1) = 1, \quad y(8) = -3, \quad I = 2^{\mathbb{N}_0} \beta(t) = 2t, t \neq 1, \beta(1) = 1.$$

Chapter 6
Higher Order Calculus of Variations

Suppose that $I \subset \mathbb{R}$ and $\beta(t) = a_1 t + a_0$ for some $a_1 \in \mathbb{R}^+$ and $a_0 \in \mathbb{R}$, for $t \in I$, is a general first kind or second kind quantum operator. Let $a, b \in I$, $a < b$.

6.1 Statement of the Variational Problem

In this chapter we investigate the variational problem.

$$\mathscr{L}(y) = \int_a^{\beta^{-1^{r-1}}(b)} L\left(t, y^{\beta^r}(t), D_\beta(y^{\beta^{r-1}})(t), \ldots, D_\beta^{r-1}(y^\beta)(t), D_\beta^r y(t)\right) D_\beta t \longrightarrow \min, \tag{6.1}$$

$$\begin{aligned} y(a) &= y_a^0, \quad y\left(\beta^{-1^{r-1}}(b)\right) = y_b^0, \\ &\vdots \\ y^{D_\beta^{r-1}}(a) &= y_a^{r-1}, \quad y^{D_\beta^{r-1}}\left(\beta^{-1^{r-1}}(b)\right) = y_b^{r-1}, \end{aligned} \tag{6.2}$$

where $y : [a, b] \to \mathbb{R}^n$, $y_a^l, y_b^l \in \mathbb{R}^n$, $l \in \{0, \ldots, r\}$, $n, r \in \mathbb{N}$.

Definition 6.1 The function L is called Lagrangian.

Assume that the Lagrangian $L(t, u_0, u_1, \ldots, u_r)$ of problem (6.1), (6.2) is continuous and has continuous partial derivatives with respect to u_0, u_1, ..., u_r, $r \geq 1$, and continuous partial delta derivative with respect to t up to order $r + 1$. For a subset $A \subseteq I$ we define

$$\mathscr{C}^{2r}(A) = \left\{ y : I \to \mathbb{R} : y^{D_\beta^{2r}} \quad is \quad continuous \quad on \quad I^{\kappa^{2r}} \right\}.$$

Let $y \in \mathscr{C}^{2r}([a, b])$.

Definition 6.2 We say that $\hat{y} \in \mathscr{C}^{2r}([a,b])$ is a (weak) local minimum of problem (6.1), (6.2) provided there exists a $D_\beta > 0$ such that

$$\mathscr{L}(\hat{y}) \leq \mathscr{L}(y)$$

for all $y \in \mathscr{C}^{2r}([a,b])$ satisfying (6.2) and

$$\|y - \hat{y}\| < D_\beta,$$

where

$$\|y\| = \sum_{l=0}^{\in fty} \left\| y^{\beta^l D_\beta^{r-l}} \right\|_{\in fty},$$

$$\|y\|_{\in fty} = \sup_{t \in [a,b]^{\kappa^r}} |y(t)|.$$

If $\mathscr{L}(\hat{y}) < \mathscr{L}(y)$ for all such $y \neq \hat{y}$ is said to be proper.

Definition 6.3 We say that $\eta \in \mathscr{C}^{2r}([a,b])$ is an admissible variation for problem (6.1), (6.2) if

$$\eta(a) = 0, \quad \eta\left(\beta^{-1^{r-1}}(b)\right) = 0,$$

$$\vdots$$

$$\eta^{D_\beta^{r-1}}(a) = 0, \quad \eta^{D_\beta^{r-1}}\left(\beta^{-1^{r-1}}(b)\right) = 0.$$

Lemma 6.1 *Let f be defined on $\left[a, \beta^{-1^{2r}}(b)\right]$ and it is continuous. Then*

$$\int_a^{\beta^{-1^{2r-1}}(b)} f(t)\eta^{\beta^r}(t) D_\beta t = 0 \tag{6.3}$$

for every admissible variation η if and only if $f(t) = 0$ for all $t \in \left[a, \beta^{-1^{2r}}(b)\right]$.

Proof If $f(t) = 0$ for all $t \in \left[a, \beta^{-1^{2r}}(b)\right]$, then the assertion is true. Let (6.3) hold. Suppose that there exists $t_0 \in \left[a, \beta^{-1^{2r}}(b)\right]$ such that $f(t_0) > 0$.

1. Let $t_0 = a$. Since f is continuous on $\left[a, \beta^{-1^{2r}}(b)\right]$, there exists a $D_\beta > 0$ such that $f(t) > 0$ for $t \in [a, a + D_\beta)$. We define η as follows.

$$\eta(t) = \begin{cases} h_{2r+2}(t,a) h_{2r+2}(a + D_\beta, t) & if \quad t \in [a, a + D_\beta) \\ 0 \quad otherwise. \end{cases}$$

We have that $\eta \in \mathscr{C}^{2r}([a,b])$ and

$$\eta(a) = \ldots = \eta^{D_\beta^{r-1}}(a) = \eta\left(\beta^{-1^{r-1}}(b)\right) = \ldots = \eta^{D_\beta^{r-1}}\left(\beta^{-1^{r-1}}(b)\right) = 0,$$

i.e., η is an admissible variation. Then

$$\begin{aligned} 0 &= \int_a^{\beta^{-1^{2r-1}}(b)} f(t)\eta^{\beta^r}(t)D_\beta t \\ &= \int_a^{a+D_\beta} f(t)\eta^{\beta^r}(t)D_\beta t \\ &> 0, \end{aligned}$$

which is a contradiction.

2. Suppose that $t_0 \neq a$ and t_0 is dense. Then there exists a $D_\beta > 0$ such that $(t_0 - D_\beta, t_0 + D_\beta) \subset [a,b]$ and $f(t) > 0$ for $t \in (t_0 - D_\beta, t_0 + D_\beta)$. Define

$$\eta(t) = \begin{cases} h_{2r+2}(t,t_0)h_{2r+2}(t_0 + D_\beta, t) & if \quad t \in (t_0 - D_\beta, t_0 + D_\beta) \\ 0 \quad otherwise. \end{cases}$$

Then

$$\begin{aligned} 0 &= \int_a^{\beta^{-1^{2r-1}}(b)} f(t)\eta^{\beta^r}(t)D_\beta t \\ &= \int_{t_0-D_\beta}^{t_0+D_\beta} f(t)\eta^{\beta^r}(t)D_\beta t \\ &> 0, \end{aligned}$$

which is a contradiction.

3. Suppose that $t_0 \neq a$ and t_0 is right-scattered. Then all points $t, t \geq t_0$ are isolated. Define η such that $\eta^{\beta^r}(t_0) = 1$ and zero elsewhere. Therefore,

$$
\begin{aligned}
0 &= \int_a^{\beta^{-1^{2r-1}}(b)} f(t)\eta^{\beta^r}(t) D_\beta t \\
&= \int_{t_0}^{\beta(t_0)} f(t)\eta^{\beta^r}(t) D_\beta t \\
&= (\beta - I)(t_0) f(t_0)\eta^{\beta^r}(t_0) \\
&> 0,
\end{aligned}
$$

which is a contradiction. This completes the proof. □

6.2 The Euler Equation

Theorem 6.1 *If $\hat{y}$ is a weak local minimum for problem* (6.1), (6.2), *then $\hat{y}$ satisfies the Euler equation*

$$\sum_{l=0}^{r} (-1)^l \left(\frac{1}{a_1}\right)^{\frac{(l-1)l}{2}} L_{u_l}^{D_\beta^l}\left(t, \hat{y}^{\beta^r}(t), \hat{y}^{\beta^{r-1} D_\beta}(t), \ldots, \hat{y}^{\beta D_\beta^{l-1}}(t), \hat{y}^{D_\beta^l}(t)\right) = 0, \quad (6.4)$$

$t \in \left[a, \beta^{-1^{2r}}(b)\right]$.

Proof Let

$$\Phi(\varepsilon) = \mathscr{L}(\hat{y} + \varepsilon\eta), \quad \varepsilon \in \mathbb{R}.$$

This function has a minimum at $\varepsilon = 0$. Therefore,

$$\Phi'(0) = 0$$

and

$$0 = \int_a^{\beta^{-1^{r-1}}(b)} \sum_{l=0}^{r} L_{u_l}\left(t, y^{\beta^r}(t), y^{\beta^{r-1}D_\beta}(t), \ldots, y^{\beta D_\beta^{r-1}}(t), y^{D_\beta^r}(t)\right)\eta^{\beta^{r-l}D_\beta^l}(t)D_\beta t$$

$$= \int_a^{\beta^{-1^r}(b)} \sum_{l=0}^{r} L_{u_l}\left(t, y^{\beta^r}(t), y^{\beta^{r-1}D_\beta}(t), \ldots, y^{\beta D_\beta^{r-1}}(t), y^{D_\beta^r}(t)\right)\eta^{\beta^{r-l}D_\beta^l}(t)D_\beta t$$

$$+ \int_{\beta^{-1^r}(b)}^{\beta^{-1^{r-1}}(b)} \sum_{l=0}^{r} L_{u_l}\left(t, y^{\beta^r}(t), y^{\beta^{r-1}D_\beta}(t), \ldots, y^{\beta D_\beta^{r-1}}(t), y^{D_\beta^r}(t)\right)\eta^{\beta^{r-l}D_\beta^l}(t)D_\beta t$$

$$= \int_a^{\beta^{-1^r}(b)} \sum_{l=0}^{r} L_{u_l}\left(t, y^{\beta^r}(t), y^{\beta^{r-1}D_\beta}(t), \ldots, y^{\beta D_\beta^{r-1}}(t), y^{D_\beta^r}(t)\right)\eta^{\beta^{r-l}D_\beta^l}(t)D_\beta t$$

$$+(\beta - I)\left(\beta^{-1^r}(b)\right)\sum_{l=0}^{r} L_{u_l}\left(\beta^{-1^r}(b), y^{\beta^r}\left(\beta^{-1^r}(b)\right), y^{\beta^{r-1}D_\beta}\left(\beta^{-1^r}(b)\right), \ldots,\right.$$

$$\left.\times y^{\beta D_\beta^{r-1}}\left(\beta^{-1^r}(b)\right), y^{D_\beta^r}\left(\beta^{-1^r}(b)\right)\right)$$

$$\times \eta^{\beta^{r-l}D_\beta^l}\left(\beta^{-1^r}(b)\right).$$

The last equality we integrate by parts and we get

$$0 = \int_a^{\beta^{-1^r}(b)} L_{u_0}\left(t, y^{\beta^r}(t), y^{\beta^{r-1}D_\beta}(t), \ldots, y^{\beta D_\beta^{r-1}}(t), y^{D_\beta^r}(t)\right)\eta^{\beta^r}(t)D_\beta t$$

$$-\sum_{l=1}^{r} \int_a^{\beta^{-1^r}(b)} L_{u_l}^{D_\beta}\left(t, y^{\beta^r}(t), y^{\beta^{r-l}D_\beta}(t), \ldots, y^{\beta D_\beta^{r-1}}(t), y^{D_\beta^r}(t)\right)\eta^{\beta^{r-1}D_\beta^{l-1}\beta}(t)D_\beta t$$

$$+\sum_{l=1}^{r} L_{u_l}\left(t, y^{\beta^r}(t), y^{\beta^{r-1}D_\beta}(t), \ldots, y^{\beta D_\beta^{r-1}}(t), y^{D_\beta^r}(t)\right)\eta^{\beta^{r-l}D_\beta^{l-1}}(t)\Big|_{t=a}^{t=\beta^{-1^r}(b)}$$

$$+(\beta - I)\left(\beta^{-1^r}(b)\right)\sum_{l=0}^{r} L_{u_l}\left(\beta^{-1^r}(b), y^{\beta^r}(\beta^{-1^r}(b)), y^{\beta^{r-1}D_\beta}\left(\beta^{-1^r}(b)\right), \ldots,\right.$$

$$\left.\times y^{\beta D_\beta^{r-1}}\left(\beta^{-1^r}(b)\right), y^{D_\beta^r}\left(\beta^{-1^r}(b)\right)\right)$$

$$\times \eta^{\beta^{r-l}D_\beta^l}\left(\beta^{-1^r}(b)\right).$$

Note that

$$\begin{aligned}
\eta^{\beta^2}(a) &= \eta^{\beta}(a) + (\beta - I)(a)\eta^{\beta D_\beta}(a) \\
&= \eta(a) + (\beta - I)(a)\eta^{D_\beta}(a) + (\beta - I)(a)a_1\eta^{D_\beta\beta}(a) \\
&= a_1(\beta - I)(a)\left(\eta^{D_\beta}(a) + (\beta - I)(a)\eta^{D_\beta^2}(a)\right) \\
&= 0, \\
\eta^{\beta^3}(a) &= \eta^{\beta^2}(a) + (\beta - I)(a)\eta^{\beta^2 D_\beta}(a) \\
&= (\beta - I)(a)a_1^2\eta^{D_\beta\beta^2}(a) \\
&= (\beta - I)(a)a_1^2\left(\eta^{D_\beta\beta}(a) + (\beta - I)(a)\eta^{D_\beta\beta D_\beta}(a)\right) \\
&= (\beta - I)(a)a_1^2\left(\eta^{D_\beta}(a) + (\beta - I)(a)\eta^{D_\beta^2}(a) + (\beta - I)(a)a_1\eta^{D_\beta^2\beta}(a)\right) \\
&= ((\beta - I)(a))^2 a_1^3\left(\eta^{D_\beta^2}(a) + (\beta - I)(a)\eta^{D_\beta^3}(a)\right) \\
&= 0,
\end{aligned}$$

and so on,

$$\eta^{\beta^{r-1}}(a) = 0,$$

and

$$\begin{aligned}
\eta^{\beta D_\beta^{r-2}}(a) &= a_1\eta^{D_\beta^{r-2}\beta}(a) \\
&= a_1\left(\eta^{D_\beta^{r-2}}(a) + (\beta - I)(a)\eta^{D_\beta^{r-1}}(a)\right) \\
&= 0, \\
\eta^{\beta^2 D_\beta^{r-3}}(a) &= a_1^2\eta^{D_\beta^{r-3}\beta^2}(a) \\
&= a_1^2\left(\eta^{D_\beta^{r-3}\beta}(a) + (\beta - I)(a)\eta^{D_\beta^{r-3}\beta D_\beta}(a)\right)
\end{aligned}$$

$$
\begin{aligned}
&= a_1^2 \left(\eta^{D_\beta^{r-3}}(a) + (\beta - I)(a)\eta^{D_\beta^{r-2}}(a) \right. \\
&\quad \left. + a_1(\beta - I)(a)\eta^{D_\beta^{r-2}\beta}(a) \right) \\
&= a_1^3(\beta - I)(a) \left(\eta^{D_\beta^{r-2}}(a) + (\beta - I)(a)\eta^{D_\beta^{r-1}}(a) \right) \\
&= 0,
\end{aligned}
$$

and so on,

$$\eta^{\beta^{r-l}D_\beta^l}(a) = 0, \quad l \in \{1, \ldots, r\}.$$

Next, note that

$$\beta^{-1}(b), \quad \beta^{-1^2}(b), \quad \ldots, \beta^{-1^{r-1}}(b)$$

are right-scattered points. We have

$$\eta^{D_\beta}\left(\beta^{-1^{r-1}}(b)\right) = 0$$

and hence,

$$0 = \frac{\eta\left(\beta^{-1^{r-1}}(b)\right) - \eta\left(\beta^{-1^{r-2}}(b)\right)}{\beta^{-1^{r-1}}(b) - \beta^{-1^{r-2}}(b)},$$

whereupon

$$\eta\left(\beta^{-1^{r-2}}(b)\right) = 0.$$

Now, using that

$$\eta^{D_\beta^2}\left(\beta^{-1^{r-1}}(b)\right) = 0,$$

we get

$$
\begin{aligned}
0 &= \frac{\eta^{D_\beta}\left(\beta^{-1^{r-1}}(b)\right) - \eta^{D_\beta}\left(\beta^{-1^{r-2}}(b)\right)}{\beta^{-1^{r-1}}(b) - \beta^{-1^{r-2}}(b)} \\
&= -\frac{\eta\left(\beta^{-1^{r-2}}(b)\right) - \eta\left(\beta^{-1^{r-3}}(b)\right)}{\left(\beta^{-1^{r-2}}(b) - \beta^{-1^{r-3}}(b)\right)\left(\beta^{-1^{r-1}}(b) - \beta^{-1^{r-2}}(b)\right)}
\end{aligned}
$$

and

$$\eta\left(\beta^{-1^{r-3}}(b)\right) = 0,$$

and so on,

$$\eta(\beta^{-1}(b)) = 0.$$

Also,

$$0 = \eta^{D_\beta^2}\left(\beta^{-1^{r-1}}(b)\right),$$

$$0 = \frac{\eta^{D_\beta}\left(\beta^{-1^{r-1}}(b)\right) - \eta^{D_\beta}\left(\beta^{-1^{r-2}}(b)\right)}{\beta^{-1^{r-1}}(b) - \beta^{-1^{r-2}}(b)}$$

$$= -\frac{\eta^{D_\beta}\left(\beta^{-1^{r-2}}(b)\right)}{\beta^{-1^{r-1}}(b) - \beta^{-1^{r-2}}(b)},$$

and

$$\eta^{D_\beta}\left(\beta^{-1^{r-2}}(b)\right) = 0$$

and so on,

$$\eta^{D_\beta}(\beta^{-1}(b)) = 0.$$

Hence,

$$0 = \eta^{\beta^{r-2}D_\beta}\left(\beta^{-1^r}(b)\right)$$

$$= a_1^{r-2}\eta^{D_\beta\beta^{r-2}}\left(\beta^{-1^r}(b)\right)$$

$$= a_1^{r-2}\eta^{D_\beta}\left(\beta^{-1^2}(b)\right)$$

$$= 0$$

and so on,

$$\eta^{\beta^{r-l}D_\beta^l}\left(\beta^{-1^r}(b)\right) = 0, \quad l \in \{1,\dots,r\}.$$

Consequently,

$$0 = \int_a^{\beta^{-1^r}(b)} L_{u_0}\left(t, y^{\beta^r}(t), y^{\beta^{r-1}D_\beta}(t), \dots, y^{\beta D_\beta^{r-1}}(t), y^{D_\beta^r}(t)\right)\eta^{\beta^r}(t)D_\beta t$$

$$-\sum_{l=1}^{r}\int_a^{\beta^{-1^r}(b)} L_{u_l}^{D_\beta}\left(t, y^{\beta^r}(t), y^{\beta^{r-l}D_\beta}(t), \dots, y^{\beta D_\beta^{r-1}}(t), y^{D_\beta^r}(t)\right)\eta^{\beta^{r-1}D_\beta^{l-1}\beta}(t)D_\beta t,$$

$$0 = \int_a^{\beta^{-1^r}(b)} \sum_{l=0}^{r} L_{u_l}\left(t, y^{\beta^r}(t), y^{\beta^{r-1}D_\beta}(t), \ldots, y^{\beta D_\beta^{r-1}}(t), y^{D_\beta^r}(t)\right)\eta^{\beta^r}(t)D_\beta t$$

$$-\sum_{l=1}^{r}\frac{1}{a_1}\int_a^{\beta^{-1^r}(b)} L_{u_l}^{D_\beta}\left(t, y^{\beta^r}(t), y^{\beta^{r-l}D_\beta}(t), \ldots, y^{\beta D_\beta^{r-1}}(t), y^{D_\beta^r}(t)\right)\eta^{\beta^{r-1+1}D_\beta^{l-1}}(t)D_\beta t.$$

The last equation we integrate by parts and so on, using Lemma 6.1, we get (6.4). This completes the proof. □

Example 6.1 Consider the variational problem

$$\mathscr{L}(y) = \int_a^{\beta^{-1}(b)} L\left(t, y^{\beta^2}(t), y^{\beta D_\beta}(t), y^{D_\beta^2}(t)\right) D_\beta t \longrightarrow \min,$$

$$y(a) = y_a^0, \quad y(\beta^{-1}(b)) = y_b^0,$$

$$y^{D_\beta}(a) = y_a^1, \quad y^{D_\beta}(\beta^{-1}(b)) = y_b^1.$$

Then the Euler equation is

$$0 = L_{u_0}\left(t, \hat{y}^{\beta^2}(t), \hat{y}^{\beta D_\beta}(t), \hat{y}^{D_\beta^2}(t)\right)$$

$$-L_{u_1}^{D_\beta}\left(t, \hat{y}^{\beta^2}(t), \hat{y}^{\beta D_\beta}(t), \hat{y}^{D_\beta^2}(t)\right)$$

$$+\frac{1}{a_1}L_{u_2}^{D_\beta^2}\left(t, \hat{y}^{\beta^2}(t), \hat{y}^{\beta D_\beta}(t), \hat{y}^{D_\beta^2}(t)\right), \quad t \in \left[a, \beta^{-1^4}(b)\right].$$

Example 6.2 Consider the variational problem

$$\mathscr{L}(y) = \int_a^{\beta^{-1^2}(b)} L\left(t, y^{\beta^3}(t), y^{\beta^2 D_\beta}(t), y^{\beta D_\beta^2}(t), y^{D_\beta^3}(t)\right) D_\beta t \longrightarrow \min,$$

$$y(a) = y_a^0, \quad y(b) = y_b^1,$$

$$y^{D_\beta}(a) = y_a^1, \quad y^{D_\beta}(b) = y_b^1,$$

$$y^{D_\beta^2}(a) = y_a^2, \quad y^{D_\beta^2}(b) = y_b^2.$$

Then the Euler equation is

$$0 = L_{u_0}\left(t, y^{\beta^3}(t), y^{\beta^2 D_\beta}(t), y^{\beta D_\beta^2}(t), y^{D_\beta^3}(t)\right)$$

$$-L_{u_1}^{D_\beta}\left(t, y^{\beta^3}(t), y^{\beta^2 D_\beta}(t), y^{\beta D_\beta^2}(t), y^{D_\beta^3}(t)\right)$$

$$+\frac{1}{a_1} L_{u_2}^{D_\beta^2}\left(t, y^{\beta^3}(t), y^{\beta^2 D_\beta}(t), y^{\beta D_\beta^2}(t), y^{D_\beta^3}(t)\right)$$

$$-\frac{1}{a_1^3} L_{u_3}^{D_\beta^3}\left(t, y^{\beta^3}(t), y^{\beta^2 D_\beta}(t), y^{\beta D_\beta^2}(t), y^{D_\beta^3}(t)\right), \quad t \in \left[a, \beta^{-1^6}(b)\right].$$

Example 6.3 Let $I = 3^{\mathbb{N}_0} \bigcup \{0\}$, $\beta(t) = 3t$, $t \in I$. Consider the variational problem

$$\mathscr{L}(y) = \int_1^{81} t y^{\beta^2}(t) y^{\beta D_\beta}(t) y^{D_\beta^2}(t) D_\beta t \longrightarrow \min,$$

$$y(1) = 0, \quad y(81) = 0,$$

$$y^{D_\beta}(1) = 1, \quad y^{D_\beta}(81) = -1.$$

Here

$$a_1 = 3, \quad r = 2,$$

$$L(t, u_0, u_1, u_2) = t u_0 u_1 u_2.$$

Then

$$L_{u_0}(t, u_0, u_1, u_2) = t u_1 u_2,$$

$$L_{u_1}(t, u_0, u_1, u_2) = t u_0 u_2,$$

$$L_{u_2}(t, u_0, u_1, u_2) = t u_0 u_1,$$

$$L_{u_1}^{D_\beta}(t, u_0, u_1, u_2) = u_0 u_2,$$

$$L_{u_2}^{D_\beta}(t, u_0, u_1, u_2) = u_0 u_1,$$

$$L_{u_2}^{D_\beta^2}(t, u_0, u_1, u_2) = 0.$$

The Euler equation is

$$ty^{\beta D_\beta}(t)y^{D_\beta^2}(t) - y^{\beta^2}(t)y^{D_\beta^2}(t) = 0, \quad t \in [1,3].$$

Exercise 6.1 Write the Euler equation for the following variational problem:

$$\int_a^{\beta^{-1^2}(b)} \left(t^2 y^{\beta^3}(t) + y^{\beta D_\beta^2} - 10 t y^{D_\beta^3}(t)\right) D_\beta t \longrightarrow \min.$$

6.3 Advanced Practical Problems

Problem 6.1 Write the Euler equation for the following variational problems:

1. $\displaystyle\int_a^{\beta^{-1^2}(b)} \left(t + y^{\beta^3}(t) + y^{D_\beta^3}(t)\right) D_\beta t \longrightarrow \min,$
2. $\displaystyle\int_a^{\beta^{-1^2}(b)} \left(t + y^{\beta^3}(t) + 2y^{\beta^2 D_\beta}(t) - 7y^{D_\beta^3}(t)\right) D_\beta t \longrightarrow \min,$
3. $\displaystyle\int_a^{\beta^{-1^3}(b)} \left(t^2 + t y^{\beta^4}(t) + t^3 y^{\beta D_\beta^3}(t)\right) D_\beta t \longrightarrow \min,$
4. $\displaystyle\int_a^{\beta^{-1}(b)} \left(t - t^3 y^{\beta^2}(t) + t^4 y^{D_\beta^2}(t)\right) D_\beta t \longrightarrow \min,$
5. $\displaystyle\int_a^{\beta^{-1^4}(b)} \left(t y^{\beta^5}(t) + y^{\beta D_\beta^4}(t) + y^{\beta^3 D_\beta^2}(t) - t^3 y^{D_\beta^5}(t)\right) d_\beta t \longrightarrow \min.$

Chapter 7
Double Integral Calculus of Variations

Let $I_1 \subset \mathbb{R}$ and $I_2 \subset \mathbb{R}$ and β_1 and β_2 be general quantum first kind or second kind quantum operators in I_1 and I_2, respectively. With $\mathscr{C}_\beta$ we denote the set of functions $f(x,y)$ on $I_1 \times I_2$ with the following properties:

1. f is continuous in x for fixed y.
2. f is continuous in y for fixed x.
3. If $(x_0,y_0) \in I_1 \times I_2$ with $x_0 = \beta_1(x_0)$ or maximal and $y_0 = \beta_2(y_0)$ or maximal, then f is continuous at (x_0,y_0).
4. If x_0 and y_0 are both left-sided, then the limit $f(x,y)$ exists (finite) as (x,y) approaches (x_0,y_0) along any path in

$$\{(x,y) \in I_1 \times I_2 : x < x_0, \quad y < y_0\}.$$

By $\mathscr{C}_\beta^{(1)}$ we denote the set of all continuous functions for which both the D_{β_1}-partial derivative and the D_{β_2}-partial derivative exist and are of the class C_β.

7.1 Statement of the Variational Problem

Let $E \subset I_1 \times I_2$ be a set of type ω and let Γ be its positively oriented fence. Suppose that a function

$$L(x,y,u,p,q), \quad (x,y) \in E \bigcup \Gamma \quad and \quad (u,p,q) \in \mathbb{R}^3,$$

is given, it is continuous together with its partial delta derivatives of the first and second order with respect to x, y and partial usual derivatives of the first and second order with respect to u, p, q. Consider the functional

$$\mathscr{L}(u) = \int\int_E L(x,y,u(\beta_1(x),\beta_2(y)),u^{D_{\beta_1}}(x,\beta_2(y)),u^{D_{\beta_2}}(\beta_1(x),y))D_{\beta_1}xD_{\beta_2}y \tag{7.1}$$

whose domain of definition $D(\mathscr{L})$ consists of functions $u \in \mathscr{C}_{\beta}^{(1)}(E\bigcup\Gamma)$ satisfying the "boundary conditions"

$$u = g(x,y) \quad on \quad \Gamma, \tag{7.2}$$

where g is a fixed function defined and continuous on Γ.

Definition 7.1 We call functions $u \in D(\mathscr{L})$ admissible.

Definition 7.2 The functions $\eta \in \mathscr{C}_{\beta}^{(1)}(E\bigcup\Gamma)$ and $\eta = 0$ on Γ are called admissible variations.

If $f \in \mathscr{C}_{\beta}^{(1)}(E\bigcup\Gamma)$, we define the norm

$$\begin{aligned} \|f\| = & \sup_{(x,y)\in E\bigcup\Gamma} |f(x,y)| + \sup_{(x,y)\in E} \left| f^{D_{\beta_1}}(x,\beta_2(y)) \right| \\ & + \sup_{(x,y)\in E} \left| f^{D_{\beta_2}}(\beta_1(x),y) \right|. \end{aligned}$$

Definition 7.3 A function $\hat{u} \in D(\mathscr{L})$ is called a weak local minimum of $\mathscr{L}$ provided there exists a $D_\beta > 0$ such that

$$\mathscr{L}(\hat{u}) \leq \mathscr{L}(u)$$

for all $u \in D(\mathscr{L})$ with

$$\|u - \hat{u}\| < D_\beta.$$

If

$$\mathscr{L}(\hat{u}) < \mathscr{L}(u)$$

for all such $u \neq \hat{u}$, then $\hat{u}$ is said to be proper weak local minimum.

7.2 First and Second Variation

For a fixed element $u \in D(\mathscr{L})$ and a fixed admissible variation η, we define Φ : $\mathbb{R} \to \mathbb{R}$ as follows.

$$\Phi(\varepsilon) = \mathscr{L}(u + \varepsilon\eta).$$

Definition 7.4 The first and second variation $\mathscr{L}$ at the point u are defined by

$$\mathscr{L}_1(u,\eta) = \Phi'(0) \quad and \quad \mathscr{L}_2(u,\eta) = \Phi''(0),$$

respectively.

Theorem 7.1 (Necessary Condition)
If $\hat{u} \in D(\mathscr{L})$ is a local minimum of $\mathscr{L}$, then

$$\mathscr{L}_1(u,\eta) = 0 \quad \textit{and} \quad \mathscr{L}_2(u,\eta) \geq 0$$

for all admissible variations η.

Proof Assume that $\mathscr{L}$ has a local minimum at $\hat{u} \in D(\mathscr{L})$. Let η be an arbitrary admissible variation. Then

$$\Phi'(0) = \mathscr{L}_1(\hat{u},\eta) \quad \textit{and} \quad \Phi''(0) = \mathscr{L}_2(\hat{u},\eta).$$

By the Taylor formula, we get

$$\Phi(\varepsilon) = \Phi(0) + \Phi'(0)\varepsilon + \frac{1}{2!}\Phi''(\alpha)\varepsilon^2,$$

where $|\alpha| \in (0, |\varepsilon|)$. If $|\varepsilon|$ is sufficiently small, then

$$\|\hat{u} + \varepsilon\eta - \hat{u}\| = |\varepsilon|\|\eta\|$$

will be as small as we please. Hence, from the definition of a local minimum, we obtain

$$\mathscr{L}(\hat{u} + \varepsilon\eta) \geq \mathscr{L}(\hat{u}),$$

i.e.,

$$\Phi(\varepsilon) \geq \Phi(0).$$

Therefore, Φ has a local minimum for $\varepsilon = 0$. From here,

$$\Phi'(0) = 0,$$

or, equivalently,

$$\mathscr{L}_1(\hat{u},\eta) = 0.$$

Since $\Phi'(0) = 0$, we have

$$\Phi(\varepsilon) - \Phi(0) = \frac{1}{2}\Phi''(\alpha)\varepsilon^2.$$

Therefore, $\Phi''(\alpha) \geq 0$ for all ε whose absolute values are sufficiently small. Letting $\varepsilon \to 0$ and using that $\alpha \to \in fty$, as $\varepsilon \to 0$, and Φ'' is continuous, we get

$$\Phi''(0) \geq 0,$$

or, equivalently,

$$\mathscr{L}_2(\hat{u},\eta) \geq 0.$$

This completes the proof. □

Theorem 7.2 (Sufficient Condition)
Let $\hat{u} \in D(\mathscr{L})$ *be such that*

$$\mathscr{L}_1(\hat{u}, \eta) = 0$$

for all admissible variations η. *If* $\mathscr{L}_2(u, \eta) \geq 0$ *for all* $u \in D(\mathscr{L})$ *and all admissible variations* η, *then* $\mathscr{L}$ *has an absolute minimum at the point* $\hat{u}$. *If* $\mathscr{L}_2(u, \eta) \geq 0$ *for all* u *in some neighborhood of the point* $\hat{u}$ *and all admissible variations* η, *then the functional* $\mathscr{L}$ *has a local minimum at* $\hat{u}$.

Proof For the function Φ we have

$$\Phi(1) = \Phi(0) + \Phi'(0) + \frac{1}{2!}\Phi''(\alpha), \quad \alpha \in (0, 1). \tag{7.3}$$

Note that

$$\begin{aligned}
\Phi(1) &= \mathscr{L}(\hat{u} + \eta), \quad \Phi(0) = \mathscr{L}(\hat{u}), \\
\Phi'(0) &= \mathscr{L}_1(\hat{u}, \eta) \\
&= 0, \\
\Phi''(\alpha) &= \left(\frac{d^2}{d\varepsilon^2}\mathscr{L}(\hat{u} + \varepsilon\eta)\right)\Big|_{\varepsilon=\alpha} \\
&= \left(\frac{d^2}{d\beta^2}\mathscr{L}(\hat{u} + \alpha\eta + \beta\eta)\right)\Big|_{\beta=0} \\
&= \mathscr{L}_2(\hat{u} + \alpha\eta, \eta).
\end{aligned}$$

From this and, we obtain

$$\mathscr{L}(\hat{u} + \eta) = \mathscr{L}(\hat{u}) + \frac{1}{2!}\mathscr{L}_2(\hat{u} + \alpha\eta, \eta)$$

for all admissible variations η, where $\alpha \in (0, 1)$. Suppose that $\mathscr{L}_2(u, \eta) \geq 0$ for all $u \in D(\mathscr{L})$ and all admissible variations η. If $u \in D(\mathscr{L})$, then putting

$$\eta = u - \hat{u},$$

we get

$$\mathscr{L}(u) \geq \mathscr{L}(\hat{u}).$$

Then $\mathscr{L}$ has an absolute minimum at the point $\hat{u}$. Now we suppose that $\mathscr{L}_2(u,\eta) \geq 0$ for all u in some neighborhood of the point $\hat{u}$ and all admissible variations η. There exists $r > 0$ such that for $u \in D(\mathscr{L})$ and

$$\|u - \hat{u}\| < r,$$

we have $\mathscr{L}_2(u,\eta) \geq 0$ for all admissible variations η. We take such an element u and we put $\eta = u - \hat{u}$. Then

$$\mathscr{L}(u) = \mathscr{L}(\hat{u}) + \frac{1}{2}\mathscr{L}_2(\hat{u} + \alpha\eta, \eta).$$

Note that

$$\begin{aligned} \|\hat{u} + \alpha\eta - \hat{u}\| &= \|\alpha\eta\| \\ &= |\alpha|\|\eta\| \\ &\leq \|\eta\| \\ &= \|u - \hat{u}\| \\ &< r. \end{aligned}$$

Hence,

$$\mathscr{L}_2(\hat{u} + \alpha\eta, \eta) \geq 0,$$

and, then

$$\mathscr{L}(u) \geq \mathscr{L}(\hat{u}).$$

This completes the proof. □

By Theorem 7.1 and Theorem 7.2, it follows that

$$\begin{aligned} \mathscr{L}_1(u,\eta) = \int\int_E \Bigg(& L_u(x,y,u(\beta_1(x),\beta_2(y)),u^{D_{\beta_1}}(x,\beta_2(y)),u^{D_{\beta_2}}(\beta_1(x),y))\eta(\beta_1(x),\beta_2(y)) \\ & + L_p(x,y,u(\beta_1(x),\beta_2(y)),u^{D_{\beta_1}}(x,\beta_2(y)),u^{D_{\beta_2}}(\beta_1(x),y))\eta^{D_{\beta_1}}(x,\beta_2(y)) \\ & + L_q(x,y,u(\beta_1(x),\beta_2(y)),u^{D_{\beta_1}}(x,\beta_2(y)),u^{D_{\beta_2}}(\beta_1(x),y))\eta^{D_{\beta_2}}(\beta_1(x),y) \Bigg) D_{\beta_1}x D_{\beta_2}y, \end{aligned} \tag{7.4}$$

and

$$\mathscr{L}_2(u,\eta) = \int\int_E \Bigg(L_{uu}(x,y,u(\beta_1(x),\beta_2(y)),u^{D_{\beta_1}}(x,\beta_2(y)),u^{D_{\beta_2}}(\beta_1(x),y))\,(\eta(\beta_1(x),\beta_2(y)))^2$$

$$+L_{pp}(x,y,u(\beta_1(x),\beta_2(y)),u^{D_{\beta_1}}(x,\beta_2(y)),u^{D_{\beta_2}}(\beta_1(x),y))\left(\eta^{D_{\beta_1}}(x,\beta_2(y))\right)^2$$

$$+L_{qq}(x,y,u(\beta_1(x),\beta_2(y)),u^{D_{\beta_1}}(x,\beta_2(y)),u^{D_{\beta_2}}(\beta_1(x),y))\left(\eta^{D_{\beta_2}}(\beta_1(x),y)\right)^2$$

$$+2L_{up}(x,y,u(\beta_1(x),\beta_2(y)),u^{D_{\beta_1}}(x,\beta_2(y)),u^{D_{\beta_2}}(\beta_1(x),y))\eta(\beta_1(x),\beta_2(y))\eta^{D_{\beta_1}}(x,\beta_2(y))$$

$$+2L_{uq}(x,y,u(\beta_1(x),\beta_2(y)),u^{D_{\beta_1}}(x,\beta_2(y)),u^{D_{\beta_2}}(\beta_1(x),y))\eta(\beta_1(x),\beta_2(y))\eta^{D_{\beta_2}}(\beta_1(x),y)$$

$$+2L_{pq}(x,y,u(\beta_1(x),\beta_2(y)),u^{D_{\beta_1}}(x,\beta_2(y)),u^{D_{\beta_2}}(\beta_1(x),y))\eta^{D_{\beta_1}}(x,\beta_2(y))\eta^{D_{\beta_2}}(\beta_1(x),y)\Bigg)$$

$$\times D_{\beta_1}x D_{\beta_2}y. \tag{7.5}$$

Example 7.1 Let

$$L(x,y,u(\beta_1(x),\beta_2(y)),u^{D_{\beta_1}}(x,\beta_2(y)),u^{D_{\beta_2}}(\beta_1(x),y))$$

$$= x+y+u(\beta_1(x),\beta_2(y)) + \left(u^{D_{\beta_1}}(x,\beta_2(y))\right)^2$$

$$+\left(u^{D_{\beta_2}}(\beta_1(x),y)\right)^3.$$

Here

$$L(x,y,u,p,q) = x+y+u+p^2+q^3.$$

Then

$$L_u(x,y,u,p,q) = 1,$$

$$L_p(x,y,u,p,q) = 2p,$$

$$L_q(x,y,u,p,q) = 3q^2,$$

$$L_{uu}(x,y,u,p,q) = 0,$$

$$L_{pp}(x,y,u,p,q) = 2,$$

$$L_{qq}(x,y,u,p,q) = 6q,$$

$$L_{uq}(x,y,u,p,q) = 0,$$

$$L_{up}(x,y,u,p,q) = 0,$$

$$L_{pq}(x,y,u,p,q) = 0.$$

Therefore, equations (7.4) and (7.5) take the form

$$\mathscr{L}_1(u,\eta) = \int\int_E \Big(\eta(\beta_1(x),\beta_2(y)) + 2u^{D_{\beta_1}}(x,\beta_2(y))\eta^{D_{\beta_1}}(x,\beta_2(y))$$

$$+3\left(u^{D_{\beta_1}}(\beta_1(x),y)\right)^2 \eta^{D_{\beta_2}}(\beta_1(x),y) \Big) D_{\beta_1}xD_{\beta_2}y,$$

$$\mathscr{L}_2(u,\eta) = \int\int_E \left(2\left(\eta^{D_{\beta_1}}(x,\beta_2(y))\right)^2 + 6u^{D_{\beta_2}}(\beta_1(x),y)\left(\eta^{D_{\beta_2}}(\beta_1(x),y)\right)^2\right)$$
$$\times D_{\beta_1}xD_{\beta_2}y.$$

Exercise 7.1 Write equations (7.4) and (7.5) for the following functionals:

1. $L(x,y,u,p,q) = xy + u^2 + upq,$
2. $L(x,y,u,p,q) = xyupq,$
3. $L(x,y,u,p,q) = x^2 + u^2p^2 + u^2q^2,$
4. $L(x,y,u,p,q) = x^2 + yupq,$
5. $L(x,y,u,p,q) = (x-yu)^2 + (y+pq)^2.$

7.3 The Euler Condition

Let E be an ω-type subset of $I_1 \times I_2$ and Γ be the positively oriented fence of E. We set

$$E^\beta = \{(x,y) \in E : (\beta_1(x),\beta_2(y)) \in E\}.$$

Lemma 7.1 (The Dubois-Reymond Lemma)

If $M(x,y)$ is continuous on $E\bigcup\Gamma$ with

$$\int\int_E M(x,y)\eta(\beta_1(x),\beta_2(y))D_{\beta_1}xD_{\beta_2}y = 0$$

for every admissible variation η, then

$$M(x,y) = 0 \quad for \quad all \quad (x,y) \in E^\beta.$$

Proof Assume the contrary. Without loss of generality, we suppose that $(x_0,y_0) \in E^\beta$ is such that $M(x_0,y_0) > 0$. The continuity of $M(x,y)$ ensures that $M(x,y)$ is positive in a rectangle

$$\Omega = [x_0,x_1) \times [y_0,y_1) \subset E$$

for some points $x_1 \in I_1$, $y_1 \in I_2$ such that

$$\beta_1(x_0) \le x_1 \quad and \quad \beta_2(y_0) \le y_1.$$

We set

$$\eta(x,y) = \begin{cases} (x-x_0)^2(x-\beta_1(x_1))^2(y-y_0)(y-\beta_2(y_1))^2 & for \quad (x,y) \in \Omega \\ 0 \quad for \quad (x,y) \in E\backslash\Omega. \end{cases}$$

We have that $\eta \in \mathscr{C}^{(1)}_\beta(E\bigcup\Gamma)$, $\eta\Big|_\Gamma = 0$, i.e., η is an admissible variation. We have that

$$\int\int_E M(x,y)\eta(\beta_1(x),\beta_2(y))D_{\beta_1}xD_{\beta_2}y = \int\int_\Omega M(x,y)\eta(\beta_1(x),\beta_2(y))D_{\beta_1}xD_{\beta_2}y$$

$$> 0,$$

which is a contradiction. This completes the proof. □

Theorem 7.3 (The Euler Necessary Condition)

Suppose that an admissible function $\hat{u}$ provides a local minimum for $\mathscr{L}$, and the function $\hat{u}$ has continuous partial delta derivatives of the second order. Then $\hat{u}$ satisfies the Euler-Lagrange equation

$$\begin{aligned} 0 &= L_u(x,y,u(\beta_1(x),\beta_2(y)),u^{D_{\beta_1}}(x,\beta_2(y)),u^{D_{\beta_2}}(\beta_1(x),y)) \\ &\quad -L_p^{D_{\beta_1}}(x,y,u(\beta_1(x),\beta_2(y)),u^{D_{\beta_1}}(x,\beta_2(y)),u^{D_{\beta_2}}(\beta_1(x),y)) \qquad (7.6) \\ &\quad -L_q^{D_{\beta_2}}(x,y,u(\beta_1(x),\beta_2(y)),u^{D_{\beta_1}}(x,\beta_2(y)),u^{D_{\beta_2}}(\beta_1(x),y)) \end{aligned}$$

for $(x,y) \in E^\beta$.

Proof Since $\hat{u}$ is a local minimum for $\mathscr{L}$, by Theorem 7.1, it follows that

$$\mathscr{L}_1(\hat{u},\eta) = 0$$

for all admissible variations η. From this and, applying integration by parts and the Green formula, we get

$$
\begin{aligned}
0 &= \mathscr{L}_1(\hat{u},\eta) \\
&= \int\int_E \Big(L_u(x,y,u(\beta_1(x),\beta_2(y)),u^{D_{\beta_1}}(x,\beta_2(y)),u^{D_{\beta_2}}(\beta_1(x),y)) \\
&\quad \times\eta(\beta_1(x),\beta_2(y))D_{\beta_1}xD_{\beta_2}y \\
&\quad +L_p(x,y,u(\beta_1(x),\beta_2(y)),u^{D_{\beta_1}}(x,\beta_2(y)),u^{D_{\beta_2}}(\beta_1(x),y)) \\
&\quad \times\eta^{D_{\beta_1}}(x,\beta_2(y)) \\
&\quad +L_q(x,y,u(\beta_1(x),\beta_2(y)),u^{D_{\beta_1}}(x,\beta_2(y)),u^{D_{\beta_2}}(\beta_1(x),y))\eta^{D_{\beta_2}}(\beta_1(x),y)\Big) \\
&\quad \times D_{\beta_1}xD_{\beta_2}y \\
&= \int\int_E L_u(x,y,u(\beta_1(x),\beta_2(y)),u^{D_{\beta_1}}(x,\beta_2(y)),u^{D_{\beta_2}}(\beta_1(x),y)) \\
&\quad \times\eta(\beta_1(x),\beta_2(y))D_{\beta_1}xD_{\beta_2}y \\
&\quad +\int\int_E \Big(L_p(x,y,u(\beta_1(x),\beta_2(y)),u^{D_{\beta_1}}(x,\beta_2(y)),u^{D_{\beta_2}}(\beta_1(x),y)) \\
&\quad \times\eta^{D_{\beta_1}}(x,\beta_2(y)) \\
&\quad +L_q(x,y,u(\beta_1(x),\beta_2(y)),u^{D_{\beta_1}}(x,\beta_2(y)),u^{D_{\beta_2}}(\beta_1(x),y)) \\
&\quad \times\eta^{D_{\beta_2}}(\beta_1(x),y)D_{\beta_1}xD_{\beta_2}y \\
&= \int\int_E L_u(x,y,u(\beta_1(x),\beta_2(y)),u^{D_{\beta_1}}(x,\beta_2(y)),u^{D_{\beta_2}}(\beta_1(x),y)) \\
&\quad \times\eta(\beta_1(x),\beta_2(y))D_{\beta_1}xD_{\beta_2}y \\
&\quad +\int\int_E \left(\frac{\partial}{D_{\beta_1}x} \Big(L_p(x,y,u(\beta_1(x),\beta_2(y)),u^{D_{\beta_1}}(x,\beta_2(y)),u^{D_{\beta_2}}(\beta_1(x),y)) \right.
\end{aligned}
$$

$$\times\eta(x,\beta_2(y))\Big)$$

$$+\frac{\partial}{D_{\beta_2}y}\Big(L_q(x,y,u(\beta_1(x),\beta_2(y)),u^{D_{\beta_1}}(x,\beta_2(y)),u^{D_{\beta_2}}(\beta_1(x),y))$$

$$\times\eta(\beta_1(x),y)\Big)\Big)D_{\beta_1}xD_{\beta_2}y$$

$$-\int\int_E\Big(L_p^{D_{\beta_1}}(x,y,u(\beta_1(x),\beta_2(y)),u^{D_{\beta_1}}(x,\beta_2(y)),u^{D_{\beta_2}}(\beta_1(x),y))$$

$$+L_q^{D_{\beta_2}}(x,y,u(\beta_1(x),\beta_2(y)),u^{D_{\beta_1}}(x,\beta_2(y)),u^{D_{\beta_2}}(\beta_1(x),y))\Big)$$

$$\times\eta(\beta_1(x),\beta_2(y))D_{\beta_1}xD_{\beta_2}y$$

$$=\int\int_E L_u(x,y,u(\beta_1(x),\beta_2(y)),u^{D_{\beta_1}}(x,\beta_2(y)),u^{D_{\beta_2}}(\beta_1(x),y))$$

$$\times\eta(\beta_1(x),\beta_2(y))D_{\beta_1}xD_{\beta_2}y$$

$$-\int\int_E\Big(\frac{\partial}{D_{\beta_1}x}\Big(L_p^{D_{\beta_1}}(x,y,u(\beta_1(x),\beta_2(y)),u^{D_{\beta_1}}(x,\beta_2(y)),u^{D_{\beta_2}}(\beta_1(x),y))$$

$$+L_q^{D_{\beta_2}}(x,y,u(\beta_1(x),\beta_2(y)),u^{D_{\beta_1}}(x,\beta_2(y)),u^{D_{\beta_2}}(\beta_1(x),y))\Big)$$

$$\times\eta(\beta_1(x),\beta_2(y))\Big)\Big)D_{\beta_1}xD_{\beta_2}y$$

$$+\int_\Gamma\Big(L_p(x,y,u(\beta_1(x),\beta_2(y)),u^{D_{\beta_1}}(x,\beta_2(y)),u^{D_{\beta_2}}(\beta_1(x),y))$$

$$\times\eta(x,\beta_2(y))D_{\beta_2}y$$

$$-L_q(x,y,u(\beta_1(x),\beta_2(y)),u^{D_{\beta_1}}(x,\beta_2(y)),u^{D_{\beta_2}}(\beta_1(x),y))$$

$$\times\eta(\beta_1(x),y)D_{\beta_1}x\Big)$$

$$= \int\int_E L_u(x,y,u(\beta_1(x),\beta_2(y)),u^{D_{\beta_1}}(x,\beta_2(y)),u^{D_{\beta_2}}(\beta_1(x),y))$$

$$\times\eta(\beta_1(x),\beta_2(y))D_{\beta_1}xD_{\beta_2}y$$

$$+\int\int_E \Big(L_p^{D_{\beta_1}}(x,y,u(\beta_1(x),\beta_2(y)),u^{D_{\beta_1}}(x,\beta_2(y)),u^{D_{\beta_2}}(\beta_1(x),y))$$

$$+L_q^{D_{\beta_2}}(x,y,u(\beta_1(x),\beta_2(y)),u^{D_{\beta_1}}(x,\beta_2(y)),u^{D_{\beta_2}}(\beta_1(x),y))\Big)$$

$$\times\eta(\beta_1(x),y)D_{\beta_1}xD_{\beta_2}y.$$

From here and from Lemma 7.1, we get (7.6). This completes the proof. □

Example 7.2 Let $I_1 = 2^{\mathbb{N}_0}\bigcup\{0\}$, $I_2 = 3^{\mathbb{N}_0}\bigcup\{0\}$, $\beta_1(t) = 2t$, $t \in I$, $\beta_2(t) = 3t$, $t \in I$. Consider the variational problem

$$\mathscr{L}(y) = \int\int_E x^2y^3u(2x,3y)u^{D_{\beta_1}}(x,3y)u^{D_{\beta_2}}(2x,y)D_{\beta_1}xD_{\beta_2}y \longrightarrow \min,$$

where

$$E = \{(x,y) \in I_1 \times I_2 : 1 \le x \le 8, \quad 1 \le y \le 27\}.$$

Here

$$L(x,y,u,p,q) = x^2y^3upq,$$

$$\beta_1(x) = 2x, \quad x \in I_1,$$

$$\beta_2(y) = 3y, \quad y \in I_2.$$

Then

$$L_u(x,y,u,p,q) = x^2y^3pq,$$

$$L_p(x,y,u,p,q) = x^2y^3uq,$$

$$L_q(x,y,u,p,q) = x^2y^3up,$$

$$L_p^{D_{\beta_1}}(x,y,u,p,q) = (\beta_1(x)+x)y^3uq$$

$$= (2x+x)y^3uq$$

$$= 3xy^3uq,$$

$$L_q^{D_{\beta_2}}(x,y,u,p,q) = x^2\left((\beta_2(y))^2 + y\beta_2(y) + y^2\right)up$$

$$= x^2\left((3y)^2 + y(3y) + y^2\right)up$$

$$= x^2(9y^2 + 3y^2 + y^2)up$$

$$= 13x^2y^2up.$$

The Euler-Lagrange equation takes the form

$$x^2y^3u^{D_{\beta_1}}(x,3y)u^{D_{\beta_2}}(2x,y) - 3xy^3u(2x,3y)u^{D_{\beta_2}}(2x,y)$$

$$-13x^2y^2u(2x,3y)u^{D_{\beta_1}}(x,3y) = 0, \quad (x,y) \in E.$$

Exercise 7.2 Write the Euler-Lagrange equation for the variational problem

$$\mathscr{L}(y) = \int\int_E \left((x^2+y^2)\,u(\beta_1(x),\beta_2(y)) + \left(u^{D_{\beta_1}}(x,\beta_2(y))\right)^2 u^{D_{\beta_2}}(\beta_1(x),y)\right) D_{\beta_1}xD_{\beta_2}y \longrightarrow \min,$$

in the cases

1. $I_1 = \mathbb{Z}$, $I_2 = 2\mathbb{Z}$, $\beta_1(t) = t+1$, $t \in I_1$, $t \neq 0$, $\beta_1(0) = 0$, $\beta_2(t) = t+2$, $t \in I_2$, $t \neq 0$, $\beta_2(0) = 0$.
2. $I_1 = I_2 = 3\mathbb{Z}$, $\beta_1(t) = \beta_2(t) = t+3$, $t \neq 0$, $\beta_1(0) = \beta_2(0) = 0$.
3. $I_1 = 2^{\mathbb{N}_0}\bigcup\{0\}$, $I_2 = 3\mathbb{Z}$, $\beta_1(t) = 2t$, $t \in I_1$, $\beta_2(t) = t+3$, $t \in I_2$, $t \neq 0$, $\beta_2(0) = 0$.
4. $I_1 = \mathbb{N}_0$, $I_2 = 3^{\mathbb{N}_0}\bigcup\{0\}$, $\beta_1(t) = t+1$, $t \in I_1$, $t \neq 0$, $\beta_1(0) = 0$, $\beta_2(t) = 3t$, $t \in I$.
5. $I_1 = I_2 = \mathbb{N}_0^3$, $\beta_1(t) = \beta_2(t)\sqrt[3]{t^3+1}$, $t \in I_1, I_2$, $t \neq 0$, $\beta_1(0) = \beta_2(0) = 0$.

7.4 Advanced Practical Problems

Problem 7.1 Write equations (7.4) and (7.5) for the following functionals:

1. $L(x,y,u,p,q) = x+p+q+u^2-(y+2p)^2$,
2. $L(x,y,u,p,q) = (x^2+p)^3+yuq$,
3. $L(x,y,u,p,q) = (x-p)^2+(y+q)^3$,
4. $L(x,y,u,p,q) = (x+y+u+p+q)^2$,
5. $L(x,y,u,p,q) = (x-u-p)^2-q^4$.

Problem 7.2 Write the Euler-Lagrange equation for the variational problem

$$\mathscr{L}(y) = \int\int_E \left((x^3-3xy+y^2+y^4)\,(u(\beta_1(x),\beta_2(y)))^2\right) D_{\beta_1}xD_{\beta_2}y \longrightarrow \min,$$

in the cases

1. $I_1 = \mathbb{Z}$, $I_2 = 2\mathbb{Z}$, $\beta_1(t) = t+1$, $t \in I_1$, $t \neq 0$, $\beta_1(0) = 0$, $\beta_2(t) = t+2$, $t \in I_2$, $t \neq 0$, $\beta_2(0) = 0$.
2. $I_1 = I_2 = 3\mathbb{Z}$, $\beta_1(t) = \beta_2(t) = t+3$, $t \neq 0$, $\beta_1(0) = \beta_2(0) = 0$.
3. $I_1 = 2^{\mathbb{N}_0} \bigcup \{0\}$, $I_2 = 3\mathbb{Z}$, $\beta_1(t) = 2t$, $t \in I_1$, $\beta_2(t) = t+3$, $t \in I_2$, $t \neq 0$, $\beta_2(0) = 0$.
4. $I_1 = \mathbb{N}_0$, $I_2 = 3^{\mathbb{N}_0} \bigcup \{0\}$, $\beta_1(t) = t+1$, $t \in I_1$, $t \neq 0$, $\beta_1(0) = 0$, $\beta_2(t) = 3t$, $t \in I$.
5. $I_1 = I_2 = \mathbb{N}_0^3$, $\beta_1(t) = \beta_2(t) \sqrt[3]{t^3+1}$, $t \in I_1, I_2$, $t \neq 0$, $\beta_1(0) = \beta_2(0) = 0$.

Chapter 8
The Noether Second Theorem

Suppose that $I \subset \mathbb{R}$ and β is a general quantum first kind or second kind quantum operator. Let also, the set I have sufficiently many points such that all computations make sense and assume that I is such that $\beta(t) = a_1 t + a_0$ for some $a_1 \in \mathbb{R}^+$ and $a_0 \in \mathbb{R}$, for any $t \in I$. Let $a, b \in I$, $a < b$, be such that all computations on $[a,b]$ make sense. For a subset $A \subseteq I$ and $k \in \mathbb{N}$ with $\mathscr{C}^k_\beta(A)$ we denote the space of all functions $y : [a,b] \to \mathbb{R}$ such that $y^{D^r_\beta}$, $r \in \{1, \ldots, k\}$, exist on A^{κ^r} and $y^{D^k_\beta}$ is rd-continuous on $[a,b]^{\kappa^k}$.

8.1 Invariance under Transformations

We will start with the following useful lemma.

Lemma 8.1 (Higher Order Fundamental Lemma of the Calculus of Variations)
Let f_0, f_1, …, $f_m \in \mathscr{C}_\beta([a,b])$. If

$$\int_a^{\beta^{-1^{m-1}}(b)} \left(\sum_{l=0}^{m} f_l(t) \eta^{\beta^{m-l} D^l_\beta}(t) \right) D_\beta t = 0,$$

for all $\eta \in \mathscr{C}^{2m}_\beta([a,b])$ such that

$$\eta(a) = 0, \quad \eta\left(\beta^{-1^{m-1}}(b)\right) = 0,$$

$$\vdots$$

$$\eta^{D^{m-1}_\beta}(a) = 0, \quad \eta^{D^{m-1}_\beta}\left(\beta^{-1^{m-1}}(b)\right) = 0,$$

then

$$\sum_{l=0}^{m}(-1)^l\left(\frac{1}{a_1}\right)^{\frac{l(l-1)}{2}} f_l^{D_\beta^l}(t)=0,\quad t\in[a,b]^{\kappa^m}.$$

Proof We will use mathematical induction.

1. Let $m=0$. Then the assertion follows from Lemma 5.1.
2. Assume that the assertion is true for some $m\in\mathbb{N}$.
3. We will prove the assertion for $m+1$. We have, using the proof of Theorem 6.1,

$$\begin{aligned}
0 &= \int_a^{\beta^{-1^m}(b)}\left(\sum_{l=0}^{m+1} f_l(t)\eta^{\beta^{m+1-l}D_{\beta^l}}(t)\right)D_\beta t\\
&= \int_a^{\beta^{-1^m}(b)}\left(\sum_{l=0}^{m} f_l(t)\eta^{\beta^{m+1-l}D_\beta^l}(t)+f_{m+1}(t)\eta^{D_\beta^{m+1}}(t)\right)d_\beta t\\
&= \int_a^{\beta^{-1^m}(b)}\left(\sum_{l=0}^{m} f_l(t)\eta^{\beta^{m+1-l}D_\beta^l}(t)\right)D_\beta t\\
&\quad+\int_a^{\beta^{-1^m}(b)}\left(f_{m+1}(t)\eta^{D_\beta^{m+1}}(t)\right)D_\beta t\\
&= \int_a^{\beta^{-1^m}(b)}\left(\sum_{l=0}^{m} f_l(t)\eta^{\beta^{m+1-l}D_\beta^l}(t)\right)D_\beta t\\
&\quad+f_{m+1}(t)\eta^{D_\beta^m}(t)\Big|_{t=0}^{t=\beta^{-1^m}(b)}\\
&\quad-\int_a^{\beta^{-1^m}(b)} f_{m+1}^{D_\beta}(t)\eta^{D_\beta^m\beta}(t)D_\beta t\\
&= \int_a^{\beta^{-1^m}(b)}\left(\sum_{l=0}^{m} f_l(t)\eta^{\beta^{m+1-l}D_\beta^l}(t)\right)D_\beta t\\
&\quad-\left(\frac{1}{a_1}\right)^m\int_a^{\beta^{-1^m}(b)} f_{m+1}^{D_\beta}(t)\eta^{\beta D_\beta^m}(t)D_\beta t\\
&= \int_a^{\beta^{-1^m}(b)}\left(\sum_{l=0}^{m-1} f_l(t)\eta^{\beta^{m+1-l}D_\beta^l}(t)+\left(f_m(t)-\left(\frac{1}{a_1}\right)^m f_{m+1}^{D_\beta}(t)\right)\eta^{\beta D_\beta^m}(t)\right)D_\beta t
\end{aligned}$$

$$
\begin{aligned}
&= \int_a^{\beta^{-1^{m-1}}(b)} \left(\sum_{l=0}^{m-1} f_l(t)\eta^{\beta^{m+1-l}D_\beta^l}(t) + \left(f_m(t) - \left(\frac{1}{a_1}\right)^m \int_a^{\beta^{-1^m}(b)} f_{m+1}^{D_\beta}(t) \right) \eta^{\beta D_\beta^m}(t) \right) D_\beta t \\
&+ \int_{\beta^{-1^m}(b)}^{\beta^{-1^{m-1}}(b)} \left(\sum_{l=0}^{m-1} f_l(t)\eta^{\beta^{m+1-l}D_\beta^l}(t) + \left(f_m(t) - \left(\frac{1}{a_1}\right)^m f_{m+1}^{D_\beta}(t) \right) \eta^{\beta D_\beta^m}(t) \right) D_\beta t \\
&= \int_a^{\beta^{-1^{m-1}}(b)} \left(\sum_{l=0}^{m-1} f_l(t)\eta^{\beta^{m+1-l}D_\beta^l}(t) + \left(f_m(t) - \left(\frac{1}{a_1}\right)^m f_{m+1}^{D_\beta}(t) \right) \eta^{\beta D_\beta^m}(t) \right) D_\beta t \\
&+(\beta - I)\left(\beta^{-1^m}(b)\right) \\
&\times \left(\sum_{l=0}^{m-1} f_l(t)\eta^{\beta^{m+1-l}D_\beta^l}(t) + \left(f_m(t) - \left(\frac{1}{a_1}\right)^m f_{m+1}^{D_\beta}(t) \right) \eta^{\beta D_\beta^m}(t) \right)\Big|_{t=\beta^{-1^m}(b)} \\
&= \int_a^{\beta^{-1^{m-1}}(b)} \left(\sum_{l=0}^{m-1} f_l(t)\eta^{\beta^{m+1-l}D_\beta^l}(t) + \left(f_m(t) - \left(\frac{1}{a_1}\right)^m f_{m+1}^{D_\beta}(t) \right) \eta^{\beta D_\beta^m}(t) \right) D_\beta t.
\end{aligned}
$$

Hence and by the induction hypothesis, we get

$$
\begin{aligned}
0 &= \sum_{l=0}^{m-1} (-1)^l \left(\frac{1}{a_1}\right)^{\frac{l(l-1)}{2}} f_l^{D_\beta^l}(t) \\
&\quad +(-1)^m \left(\frac{1}{a_1}\right)^{\frac{m(m-1)}{2}} \left(f_m(t) - \left(\frac{1}{a_1}\right)^m f_{m+1}^{D_\beta}(t) \right) \\
&= \sum_{l=0}^{m+1} (-1)^l \left(\frac{1}{a_1}\right)^{\frac{m(m+1)}{2}} f_l^{D_\beta^l}(t), \quad t \in [a,b]^{\kappa^{m+1}}.
\end{aligned}
$$

This completes the proof. □

Suppose that

$$y = (y_1, y_2, \ldots, y_n).$$

We consider the transformations

$$
\begin{aligned}
\bar{t} &= t, \\
\bar{y}_{k+1}(t) &= y_k(t) + \sum_{j=1}^{k} T^{kj}(p_j)(t), \quad k \in \{1, \ldots, n\},
\end{aligned}
\tag{8.1}
$$

where

$$T^{kj}(p_j) = \sum_{l=0}^{m} g_{lj}^k p_j^{\beta^{m-(l+1)}D_\beta^l}$$

and $g_{lj}^k \in \mathscr{C}_\beta^1([a,b])$, $g_{lj}^k : [a,b] \to \mathbb{R}$, $l \in \{0,\ldots,m\}$, $j \in \{1,\ldots,r\}$, $k \in \{1,\ldots,n\}$, which depend on arbitrary functions $p_1, p_2, \ldots, p_r \in \mathscr{C}_\beta^m([a,\beta^m(b)])$, $p_l : [a,b] \to \mathbb{R}$, $l \in \{1,\ldots,r\}$ and their delta derivatives up to order m. Consider the variational problem

$$\mathscr{L}(y) = \int_a^b L\left(t, y^\beta(t), y^{D_\beta}(t)\right) D_\beta t \longrightarrow \quad extremize, \tag{8.2}$$

where $L : [a,b] \times \mathbb{R}^{2n} \to \mathbb{R}$, $L = L(t,u,v)$, is a given function that is continuous and has continuous partial derivatives with respect to u and v, and continuous partial delta derivative with respect to t.

Definition 8.1 The functional $\mathscr{L}$ is called invariant under the transformations (8.1) if and only if

$$\int_a^b L\left(t, y^\beta(t), y^{D_\beta}(t)\right) D_\beta t = \int_a^b L\left(\bar{t}, \bar{y}^\beta(t), \bar{y}^{D_\beta}(t)\right) D_\beta t.$$

Theorem 8.1 (Necessary Condition of Invariance)

If the functional $\mathscr{L}$ is invariant under the transformations (8.1), *then*

$$\begin{aligned} 0 = \sum_{k=1}^n \int_a^b \Bigg(& L_{y_k^\beta}\left(t, y^\beta(t), y^{D_\beta}(t)\right) \left(\sum_{j=1}^r T^{kj}(p_j)\right)^\beta (t) \\ & + L_{y_k^{D_\beta}}\left(t, y^\beta(t), y^{D_\beta}(t)\right) \left(\sum_{j=1}^r T^{kj}(p_j)\right)^{D_\beta} (t) \Bigg) D_\beta t. \end{aligned} \tag{8.3}$$

Proof Since $\mathscr{L}$ is invariant under the transformations (8.1), we have

$$\begin{aligned} & \int_a^b L\left(t, y^\beta(t), y^{D_\beta}(t)\right) D_\beta t \\ & = \int_a^b L\Bigg(t, y_1^\beta(t) + \varepsilon \left(\sum_{j=1}^r T^{1j}(p_j)\right)^\beta (t), \ldots, \\ & y_n^\beta(t) + \varepsilon \left(\sum_{j=1}^r T^{nj}(p_j)\right)^\beta (t), \\ & y_1^{D_\beta}(t) + \varepsilon \left(\sum_{j=1}^r T^{1j}(p_j)\right)^{D_\beta} (t), \\ & y_n^{D_\beta}(t) + \varepsilon \left(\sum_{j=1}^r T^{nj}(p_j)\right)^{D_\beta} (t) \Bigg) D_\beta t \end{aligned}$$

for any $\varepsilon \in \mathbb{R}$. Differentiating the last equality with respect to ε and then taking $\varepsilon = 0$, we get (8.3). This completes the proof. □

8.2 The Noether Second Theorem without Transformations of Time

Define

$$E_k(L) = L^{\beta}_{y_k} - L^{D_\beta}_{y_k^{D_\beta}}, \quad k = 1, \ldots, n,$$

where

$$L^{D_\beta}_{y_k^{D_\beta}} = \frac{D_\beta}{D_\beta t} L_{y_k^{D_\beta}}.$$

Theorem 8.2 *If the functional* $\mathscr{L}$ *is invariant under the transformations* (8.1), *then*

$$\sum_{k=1}^{n} \sum_{l=0}^{m} (-1)^l \left(\frac{1}{a_1}\right)^{\frac{l(l+1)}{2}} \left(\left(g^k_{lj}\right)^{\beta} E_k(L)\right)^{D^l_\beta} = 0.$$

Proof By Theorem 8.1, we have

$$0 = \sum_{k=1}^{n} \int_a^b \left(L_{y_k^\beta}\left(t, y^\beta(t), y^{D_\beta}(t)\right) \left(\sum_{j=1}^{r} T^{kj}(p_j)\right)^{\beta} (t) \right.$$

$$\left. + L_{y_k^{D_\beta}}\left(t, y^\beta(t), y^{D_\beta}(t)\right) \left(\sum_{j=1}^{r} T^{kj}(p_j)\right)^{D_\beta} (t) \right) D_\beta t.$$

We fix $j \in \{1, \ldots, r\}$. By the arbitrariness of the functions p_1, p_2, …, p_r, we can suppose that $p_h \equiv 0$ for $h \neq j$, $h \in \{1, \ldots, r\}$, and

$$p_j(a) = 0, \quad p_j(b) = 0,$$

$$\vdots$$

$$p_j^{D^{m-1}_\beta}(a) = 0, \quad p_j^{D^{m-1}_\beta}(b) = 0,$$

$$p_j^{\beta^{-1} D^m_\beta}(a) = 0, \quad p_j^{\beta^{-1} D^m_\beta}(b) = 0.$$

Therefore,

$$0=\sum_{k=1}^{n}\int_a^b\Big(L_{y_k^\beta}\left(t,y^\beta(t),y^{D_\beta}(t)\right)\left(T^{kj}(p_j)\right)^\beta(t)$$

$$+L_{y_k^{D_\beta}}\left(t,y^\beta(t),y^{D_\beta}(t)\right)\left(T^{kj}(p_j)\right)^{D_\beta}(t)\Big)D_\beta t.$$

Integrating by parts the last equality, we get

$$0=\sum_{k=1}^{n}\Bigg(\int_a^b\left(L_{y_k^\beta}\left(t,y^\beta(t),y^{D_\beta}(t)\right)-L^{D_\beta}_{y_k^{D_\beta}}\left(t,y^\beta(t),y^{D_\beta}(t)\right)\right)$$

$$\times\left(T^{kj}(p_j)\right)^\beta(t)D_\beta t$$

$$+L_{y_k}\left(t,y^\beta(t),y^{D_\beta}(t)\right)T^{kj}(p_j)(t)\Big|_{t=a}^{t=b}\Bigg).$$

By the proof of Theorem 6.1, we obtain

$$T^{kj}(p_j)(t)\Big|_{t=a}^{t=b}=0,\quad k\in\{1,\dots,n\}.$$

Therefore,

$$0=\sum_{k=1}^{n}\int_a^b\left(L_{y_k^\beta}\left(t,y^\beta(t),y^{D_\beta}(t)\right)-L^{D_\beta}_{y_k^{D_\beta}}\left(t,y^\beta(t),y^{D_\beta}(t)\right)\right)$$

$$\times\left(T^{kj}(p_j)\right)^\beta(t)D_\beta t,$$

or

$$\sum_{k=1}^{n}\int_a^b E_k(L)\left(t,y^\beta(t),y^{D_\beta}(t)\right)\left(T^{kj}(p_j)\right)^\beta(t)D_\beta t=0,$$

or

$$0=\sum_{k=1}^{n}\int_a^b E_k(L)\left(t,y^\beta(t),y^{D_\beta}(t)\right)$$

$$\times\left(\sum_{l=0}^{m}g_{lj}^k p_j^{\beta^{m-(l+1)}D_\beta^l}\right)^\beta(t)D_\beta t,$$

or

$$0 = \int_a^b \sum_{l=0}^{m} \sum_{k=1}^{n} E_k(L)\left(t, y^{\beta}(t), y^{D_\beta}(t)\right)$$

$$\times \left(g_{lj}^k\right)^{\beta} \left(\frac{1}{a_1}\right)^{l} p_j^{\beta^{m-l} D_\beta^l}(t) D_\beta t.$$

From this and, we obtain

$$\sum_{l=0}^{m} \sum_{k=1}^{n} (-1)^l \left(\frac{1}{a_1}\right)^{\frac{l(l-1)}{2}} \left(E_k(L)\left(g_{lj}^k\right)^{\beta}\left(\frac{1}{a_1}\right)^{l}\right)^{D_\beta^l} = 0,$$

or

$$\sum_{k=1}^{n} \sum_{l=0}^{m} (-1)^l \left(\frac{1}{a_1}\right)^{\frac{l(l+1)}{2}} \left(\left(g_{lj}^k\right)^{\beta} E_k(L)\right)^{D_\beta^l} = 0.$$

This completes the proof. □

8.3 The Noether Second Theorem with Transformations of Time

In this section we suppose that the Lagrangian L is defined for all $t \in \mathbb{R}$, not only for t from the set I, and L is continuous and has continuous partial derivative with respect to t. Consider the transformations

$$\begin{aligned} \bar{t} &= t + \sum_{j=1}^{r} H^j(p_j)(t), \\ \bar{y}_k(\bar{t}) &= y_k(t) + \sum_{j=1}^{r} T^{kj}(p_j)(t), \quad k \in \{1, \ldots, n\}, \end{aligned} \tag{8.4}$$

where

$$H^j(p_j) = \sum_{l=0}^{m} f_{lj} p_j^{\beta^{m-(l+1)} D_\beta^l},$$

$$T^{kj}(p_j) = \sum_{l=0}^{m} g_{lj}^k p_j^{\beta^{m-(l+1)} D_\beta^l},$$

$f_{lj}, g_{lj}^k \in \mathscr{C}_\beta^1([a,b])$, $k \in \{1, \ldots, n\}$, $l \in \{0, \ldots, m\}$, $j \in \{1, \ldots, r\}$, which depend on arbitrary functions $p_1, \ldots, p_r$, $p_s \in \mathscr{C}_\beta^{2m}([a, \beta^m(b)])$, and their partial derivatives up to order m. Assume that the map

$$t \longmapsto \alpha(t) = t + \sum_{j=1}^{r} H^j(p_j)(t)$$

is strictly increasing $\mathscr{C}^1_\beta$ function and its image is again a set $\overline{I}$ with general first kind or second kind quantum operator $\overline{\beta}$.

Definition 8.2 The functional $\mathscr{L}$ is called invariant under the transformations (8.4) if and only if for all $y \in \mathscr{C}^1_\beta([a,b])$ we have

$$\int_a^b L\left(t, y^\beta(t), y^{D_\beta}(t)\right) D_\beta t = \int_{\overline{a}}^{\overline{b}} L\left(\overline{t}, \overline{y}^{\overline{\beta}}(\overline{t}), \overline{y}^{\overline{D_\beta}}(\overline{t})\right) \overline{D_\beta \overline{t}}.$$

Theorem 8.3 (The Noether Second Theorem with Transformations of Time) *If $\mathscr{L}$ is invariant under the transformations* (8.4), *then*

$$\begin{aligned} 0 = \sum_{k=1}^{n} \sum_{l=1}^{m} (-1)^l \left(\frac{1}{a_1}\right)^{\frac{l(l+1)}{2}} \Bigg(\left(\left(g^k_{lj}\right)^\beta E_k(L)\right)^{D^l_\beta} \\ + \left((f_{lj})^\beta \left(L_t - \left(L - y_k^{D_\beta} L_{y_k^{D_\beta}} - (\beta - I) L_t \right)^{D_\beta} \right)\right)^{D^l_\beta} \Bigg) \end{aligned} \tag{8.5}$$

for $j \in \{1, 2, \ldots, r\}$.

Proof Let $q \neq 0$. We define

$$\widetilde{L}(t, s, y, q, v) = L\left(s - (\beta(t) - t) q, y, \frac{v}{q}\right) q.$$

For $s(t) = t$ and for any $y \in \mathscr{C}^1_\beta([a,b])$, $y : [a,b] \to \mathbb{R}^n$, we have

$$L\left(t, y^\beta(t), y^{D_\beta}(t)\right) = \widetilde{L}\left(t, s^\beta(t), y^\beta(t), s^{D_\beta}(t), y^{D_\beta}(t)\right).$$

Therefore, for $s(t) = t$, we get

$$\begin{aligned} \mathscr{L}(y) &= \int_a^b L\left(t, y^\beta(t), y^{D_\beta}(t)\right) D_\beta t \\ &= \int_a^b \widetilde{L}\left(t, s^\beta(t), y^\beta(t), s^{D_\beta}(t), y^{D_\beta}(t)\right) D_\beta t \\ &= \mathscr{L}(s, y) \end{aligned}$$

and using that $\mathscr{L}$ is invariant under the transformations (8.4), we get

$$\begin{aligned}
\mathscr{L}(s,y) &= \int_a^b L\left(t, y^\beta(t), y^{D_\beta}(t)\right) D_\beta t \\
&= \int_{\alpha(a)}^{\alpha(b)} L\left(\bar{t}, \left(\bar{y}\circ\bar{\beta}\right)(\bar{t}), \bar{y}^{\overline{D_\beta}}(\bar{t})\right) \overline{D_\beta \bar{t}} \\
&= \int_a^b L\left(\alpha(t), \left(\bar{y}\circ\bar{\beta}\circ\alpha\right)(t), \bar{y}^{\overline{D_\beta}}(\alpha(t))\right) \alpha^{D_\beta}(t) D_\beta t \\
&= \int_a^b L\left(\alpha(t), (\bar{y}\circ\alpha\circ\beta)(t), \frac{(\bar{y}\circ\alpha)^{D_\beta}(t)}{\alpha^{D_\beta}(t)}\right) \alpha^{D_\beta}(t) D_\beta t \\
&= \int_a^b L\left(\alpha^\beta(t) - (\beta(t)-t)\alpha^{D_\beta}(t), (\bar{y}\circ\alpha)^\beta(t), \frac{(\bar{y}\circ\alpha)^{D_\beta}(t)}{\alpha^{D_\beta}(t)}\right) \alpha^{D_\beta}(t) D_\beta t \\
&= \int_a^b \widetilde{L}\left(t, \alpha^\beta(t), (\bar{y}\circ\alpha)^\beta(t), \alpha^{D_\beta}(t), (\bar{y}\circ\alpha)^{D_\beta}(t)\right) D_\beta t \\
&= \widetilde{\mathscr{L}}(\alpha, \bar{y}\circ\alpha).
\end{aligned}$$

Let

$$H(t,y(t)) = \alpha(t),$$

$$T^k(t,y(t)) = y_k(t) + \sum_{j=1}^{r} T^{kj}(p_j)(t), \quad k \in \{1,\ldots,n\},$$

$$T = \left(T^1, T^2, \ldots, T^n\right).$$

Hence, for $s(t) = t$, we have

$$\begin{aligned}
(\alpha(t), (\bar{y}\circ\alpha)(t)) &= (\bar{t}, \bar{y}(\bar{t})) \\
&= (H(t,y(t)), T(t,y(t))) \\
&= (H(s(t),y(t)), T(s(t),y(t))).
\end{aligned}$$

Therefore, for $s(t) = t$, we obtain

$$\widetilde{\mathscr{L}}(s,y) = \widetilde{\mathscr{L}}(H(s,y),T(s,y)).$$

Consequently $\widetilde{\mathscr{L}}$ is invariant on

$$\widetilde{U} = \left\{(s,y) : s(t) = t, \quad y \in \mathscr{C}^1_\beta([a,b])\right\}$$

under the group of state transformations

$$(\bar{s},\bar{y}) = (H(s,y),T(s,y))$$

in the sense of Definition 8.1. From this and, we get

$$\begin{aligned} 0 &= \sum_{k=1}^{n}\sum_{l=0}^{m}(-1)^l\left(\frac{1}{a_1}\right)^{\frac{l(l+1)}{2}}\left(\left(g^k_{lj}\right)^\beta E_k\left(\tilde{L}\right)\right)^{D^l_\beta} \\ &\quad +\sum_{l=0}^{m}(-1)^l\left(\frac{1}{a_1}\right)^{\frac{l(l+1)}{2}}\left((f_{lj})^\beta E_s(\widetilde{L})\right)^{D^l_\beta}, \end{aligned} \tag{8.6}$$

where

$$E_s(\widetilde{L}) = \widetilde{L}_{s^\beta} - \widetilde{L}^{D_\beta}_{s^{D_\beta}}.$$

Note that

$$\begin{aligned} &\widetilde{L}_{s^\beta}\left(t, s^\beta(t), y^\beta(t), s^{D_\beta}(t), y^{D_\beta}(t)\right) \\ &= L_t\left(s^\beta(t) - (\beta(t)-t)s^{D_\beta}(t), y^\beta(t), \frac{y^{D_\beta}(t)}{s^{D_\beta}(t)}\right)s^{D_\beta}(t) \end{aligned}$$

and

$$\begin{aligned} &\widetilde{L}_{s^{D_\beta}}\left(t, s^\beta(t), y^\beta(t), s^{D_\beta}(t), y^{D_\beta}(t)\right) \\ &= L\left(s^\beta(t) - (\beta(t)-t)s^{D_\beta}(t), y^\beta(t), \frac{y^{D_\beta}(t)}{s^{D_\beta}(t)}\right) \\ &\quad -\sum_{k=1}^{n}\frac{y_k^{D_\beta}(t)}{s^{D_\beta}(t)}L_{y_k^{D_\beta}}\left(s^\beta(t) - (\beta(t)-t)s^{D_\beta}(t), y^\beta(t), \frac{y^{D_\beta}(t)}{s^{D_\beta}(t)}\right) \\ &\quad -L_t\left(s^\beta(t) - (\beta(t)-t)s^{D_\beta}(t), y^\beta(t), \frac{y^{D_\beta}(t)}{s^{D_\beta}(t)}\right)(\beta(t)-t)s^{D_\beta}(t). \end{aligned}$$

Hence, for $s(t) = t$, we obtain

$$\begin{aligned}
&E_s(\widetilde{L})\left(t, s^{\beta}(t), y^{\beta}(t), s^{D_\beta}(t), y^{D_\beta}(t)\right) \\
&= L_t\left(t, y^{\beta}(t), y^{D_\beta}(t)\right) \\
&-\left(L\left(t, y^{\beta}(t), y^{D_\beta}(t)\right) - \sum_{k=1}^{n} y_k^{D_\beta}(t) L_{y_k^{D_\beta}}\left(t, y^{\beta}(t), y^{D_\beta}(t)\right)\right. \\
&\left. -(\beta(t)-t) L_t\left(t, y^{\beta}(t), y^{D_\beta}(t)\right)\right)^{D_\beta},
\end{aligned} \tag{8.7}$$

$$E_k(\widetilde{L})\left(t, s^{\beta}(t), y^{\beta}(t), s^{D_\beta}(t), y^{D_\beta}(t)\right) = E_k(L)\left(t, y^{\beta}(t), y^{D_\beta}(t)\right) \tag{8.8}$$

for $k \in \{1, \ldots, n\}$. We substitute (8.7) and (8.8) into (8.6) and we get (8.5). This completes the proof. □

8.4 The Noether Second Theorem-Double Delta Integral Case

Let $I_1 \subset \mathbb{R}$ and $I_2 \subset \mathbb{R}$ and β_1 and β_2 be general quantum first kind or second kind quantum operators in I_1 and I_2, respectively. Let $E \subseteq I_1 \times I_2$ be an ω-type set and let Γ be its positively oriented fence and

$$E^{\beta} = \{(x,y) \in E : (\beta_1(x), \beta_2(y)) \in E\}.$$

Suppose that $L(x,y,u,p,q)$, $(x,y) \in E \bigcup \Gamma$, $(u,p,q) \in \mathbb{R}^{3n}$, are (or is) given. Let also, L be continuous together with its partial delta derivatives of first and second order with respect to x and y, and partial usual derivatives of the first and second order with respect to u, p, q. Consider the variational problem

$$\begin{aligned}
\mathscr{L}(u) &= \int\int_E L\left(x, y, u(\beta_1(x), \beta_2(y)), u^{D_{\beta_1}}(x, \beta_2(y)), u^{D_{\beta_2}}(\beta_1(x), y)\right) D_{\beta_1} x D_{\beta_2} y \\
&\longrightarrow \textit{extremize},
\end{aligned}$$

where the set of admissible functions is

$$D = \left\{u : E \bigcup \Gamma \to \mathbb{R}^n, \quad u \in \mathscr{C}_{\beta}^{(1)}, \quad u = g \quad on \quad \Gamma\right\},$$

g is a fixed function defined and continuous on Γ. With ${\beta^{-1}}_1$ and ${\beta^{-1}}_2$ we denote the backward jump operators of I_1 and I_2, respectively. Suppose that I_1 and I_2 are such that

$$\beta_1(\beta^{-1}{}_1(x)) = x, \quad x \in I_{1\kappa},$$

$$\beta_2(\beta^{-1}{}_2(y)) = y, \quad y \in I_{2\kappa}.$$

Let

$$u(x,y) = (u_1(x,y), u_2(x,y), \dots, u_n(x,y)).$$

Consider the transformations

$$\begin{aligned} \overline{x} &= x, \\ \overline{y} &= y, \\ \overline{u}_k(\overline{x},\overline{y}) &= u_k(x,y) + T^k(p)(x,y), \quad k \in \{1,\dots,n\}, \end{aligned} \tag{8.9}$$

where

$$\begin{aligned} T^k(p)(x,y) = {} & a_0^k(x,y)p(x,y) \\ & + a_1^k(x,y)p^{D_{\beta_1}}(x,y) \\ & + a_2^k(x,y)p^{D_{\beta_2}}(x,y), \quad k \in \{1,\dots,n\}, \end{aligned}$$

where a_0, a_1 and a_2 are $\mathscr{C}^1$ functions and p has continuous partial delta derivatives of the first and second order. Note that the transformations (8.9) depend on an arbitrary continuous function p and the partial derivatives of p.

Definition 8.3 The functional $\mathscr{L}$ is called invariant under the transformations (8.9) if and only if

$$\begin{aligned} & \int\int_E L\left(x, y, u(\beta_1(x), \beta_2(y)), u^{D_{\beta_1}}(x, \beta_2(y)), u^{D_{\beta_2}}(\beta_1(x), y)\right) D_{\beta_1}x D_{\beta_2}y \\ & = \int\int_E L\left(x, y, \overline{u}(\beta_1(x), \beta_2(y)), \overline{u}^{D_{\beta_1}}(x, \beta_2(y)), \overline{u}^{D_{\beta_2}}(\beta_1(x), y)\right) D_{\beta_1}x D_{\beta_2}y. \end{aligned}$$

We introduce the following notations:

$$\begin{aligned} T^k(p^\beta)(x,y) &= T^k(\beta_1(x), \beta_2(y)), \\ T^k(p^{\beta_1})(x,y) &= T^k(\beta_1(x), y), \\ T^k(p^{\beta_2})(x,y) &= T^k(x, \beta_2(y)), \quad k \in \{1,\dots,n\}, \quad x \in I_1, \quad y \in I_2. \end{aligned}$$

Theorem 8.4 (Necessary Condition of Invariance) *If thefunctional $\mathscr{L}$ is invariant under the transformations* (8.9), *then*

$$0 = \sum_{k=1}^{n} \int\int_{E} \left(L_{u_k^{\beta}} T^k(p^{\beta}) + L_{u_k^{D_{\beta_1}}} T^{kD_{\beta_1}}\left(p^{\beta_2}\right) \right. \tag{8.10}$$
$$\left. + L_{u_k^{D_{\beta_2}}} T^{kD_{\beta_2}}\left(p^{\beta_1}\right) \right) D_{\beta_1}x D_{\beta_2}y.$$

Proof Let $\varepsilon \in \mathbb{R}$. Since $\mathscr{L}$ is invariant under the transformations (8.9), we have

$$\int\int_{E} L\left(x, y, u(\beta_1(x), \beta_2(y)), u^{D_{\beta_1}}(x, \beta_2(y)), u^{D_{\beta_2}}(\beta_1(x), y)\right) D_{\beta_1}x D_{\beta_2}y$$
$$= \int\int_{E} L\left(x, y, u_1(x,y) + \varepsilon T^1(p)(x,y), \ldots, u_n(x,y) + \varepsilon T^n(p)(x,y), \right.$$
$$u_1^{D_{\beta_1}}(x, \beta_2(y)) + \varepsilon T^{1D_{\beta_1}}(p)(x, \beta_2(y)), \ldots, u_n^{D_{\beta_1}}(x, \beta_2(y)) + \varepsilon T^{nD_{\beta_1}}(p)(x,y),$$
$$\left. u_1^{D_{\beta_2}}(\beta_1(x), y) + \varepsilon T^{1D_{\beta_2}}(p)(x, \beta_2(y)), \ldots, u_n^{D_{\beta_2}}(\beta_1(x), y) + \varepsilon T^{nD_{\beta_2}}(p)(x,y) \right)$$
$$\times D_{\beta_1}x D_{\beta_2}y.$$

We differentiate the last equality with respect to ε and then letting $\varepsilon \to 0$, we get (8.10). This completes the proof. □

Define

$$\hat{E}_k(L) = L_{u_k^{\beta}} - L_{u_k^{D_{\beta_1}}}^{D_{\beta_1}} - L_{u_k^{D_{\beta_2}}}^{D_{\beta_2}}, \quad k \in \{1, \ldots, n\},$$

where

$$L_{u_k^{D_{\beta_1}}}^{D_{\beta_1}} = \frac{\partial}{D_{\beta_1}} L_{u_k^{D_{\beta_1}}}, \quad L_{u_k^{D_{\beta_2}}}^{D_{\beta_2}} = \frac{\partial}{D_{\beta_2}} L_{u_k^{D_{\beta_2}}}.$$

Theorem 8.5 *If $\mathscr{L}$ is invariant under the transformations* (8.9), *then*

$$\sum_{k=1}^{n} \int\int_{E} \hat{E}_k(L) T^k(p^{\beta}) D_{\beta_1}x D_{\beta_2}y = 0. \tag{8.11}$$

Proof We take p such that

$$p(x, \beta_2(y))\Big|_{\Gamma} = 0,$$

$$p(\beta_1(x), y)\Big|_{\Gamma} = 0,$$

$$p^{D_{\beta_1}}(\beta^{-1}{}_1(x),\beta_2(y))\Big|_{\Gamma}=0,$$

$$p^{D_{\beta_1}}(x,y)\Big|_{\Gamma}=0,$$

$$p^{D_{\beta_2}}(x,y)\Big|_{\Gamma}=0,$$

$$p^{D_{\beta_2}}(\beta_1(x),\beta^{-1}{}_2(y))\Big|_{\Gamma}=0.$$

We fix $l\in\{1,\ldots,n\}$. Then, using the Green formula, we get

$$\begin{aligned}
&\int\int_E\left(L_{u_l^{D_{\beta_1}}}T^{lD_{\beta_1}}(p^{\beta_2})+L_{u_l^{D_{\beta_2}}}T^{lD_{\beta_2}}(p^{\beta_1})\right)D_{\beta_1}xD_{\beta_2}y\\
&=\int\int_E\left(\left(L_{u_l^{D_{\beta_1}}}T^{l}(p^{\beta_2})\right)^{D_{\beta_1}}+\left(L_{u_l^{D_{\beta_2}}}T^{l}(p^{\beta_1})\right)^{D_{\beta_2}}\right)D_{\beta_1}xD_{\beta_2}y\\
&\quad-\int\int_E\left(L^{D_{\beta_1}}_{u_l^{D_{\beta_1}}}T^{l}(p^{\beta})+L^{D_{\beta_2}}_{u_l^{D_{\beta_2}}}T^{l}(p^{\beta})\right)D_{\beta_1}xD_{\beta_2}y\\
&=\int_{\Gamma}\left(L_{u_l^{D_{\beta_1}}}T^{l}(p^{\beta_2})D_{\beta_2}y-L_{u_l^{D_{\beta_2}}}T^{l}(p^{\beta_1})D_{\beta_1}x\right)\\
&\quad-\int\int_E\left(L^{D_{\beta_1}}_{u_l^{D_{\beta_1}}}T^{l}(p^{\beta})+L^{D_{\beta_2}}_{u_l^{D_{\beta_2}}}T^{l}(p^{\beta})\right)D_{\beta_1}xD_{\beta_2}y\\
&=-\int\int_E\left(L^{D_{\beta_1}}_{u_l^{D_{\beta_1}}}T^{l}(p^{\beta})+L^{D_{\beta_2}}_{u_l^{D_{\beta_2}}}T^{l}(p^{\beta})\right)D_{\beta_1}xD_{\beta_2}y.
\end{aligned}$$

Hence,

$$\begin{aligned}
&\sum_{l=1}^{n}\int\int_E\left(L_{u_l^{D_{\beta_1}}}T^{lD_{\beta_1}}(p^{\beta_2})+L_{u_l^{D_{\beta_2}}}T^{lD_{\beta_2}}(p^{\beta_1})\right)D_{\beta_1}xD_{\beta_2}y\\
&=-\sum_{l=1}^{n}\int\int_E\left(L^{D_{\beta_1}}_{u_l^{D_{\beta_1}}}T^{l}(p^{\beta})+L^{D_{\beta_2}}_{u_l^{D_{\beta_2}}}T^{l}(p^{\beta})\right)D_{\beta_1}xD_{\beta_2}y.
\end{aligned}\tag{8.12}$$

By Theorem 8.4, we have that

$$\begin{aligned}
\sum_{k=1}^{n}\int\int_E L_{u_k^{\beta}}T^{k}(p^{\beta})D_{\beta_1}xD_{\beta_2}y=-\sum_{k=1}^{n}\int\int_E&\Bigg(L_{u_k^{D_{\beta_1}}}T^{kD_{\beta_1}}(p^{\beta})\\
&+L_{u_k^{D_{\beta_2}}}T^{kD_{\beta_2}}(p^{\beta})\Bigg)D_{\beta_1}xD_{\beta_2}y.
\end{aligned}$$

Hence and by (8.12), we get (8.11). This completes the proof. □

Theorem 8.6 *For any* $k \in \{1,\dots,n\}$, *we have*

$$\int\int_E qT^k(p^\beta)D_{\beta_1}xD_{\beta_2}y = \int\int_E \Big(qa_0^k - (qa_1^k)^{D_{\beta_1}} - (qa_2^k)^{D_{\beta_2}} \Big) p^\beta D_{\beta_1}xD_{\beta_2}y.$$

Proof We take p such that

$$p(\beta_1(x),y)\Big|_\Gamma = 0, \quad p(x,\beta_2(y))\Big|_\Gamma = 0.$$

Note that

$$\begin{aligned}\int\int_E qT^k(p^\beta)D_{\beta_1}xD_{\beta_2}y &= \int\int_E qa_0^k(x,y)p(\beta_1(x),\beta_2(y))D_{\beta_1}xD_{\beta_2}y \\ &+\int\int_E qa_1^k(x,y)p^{D_{\beta_1}}(x,\beta_2(y))D_{\beta_1}xD_{\beta_2}y \\ &+\int\int_E qa_2^k(x,y)p^{D_{\beta_2}}(\beta_1(x),y)D_{\beta_1}xD_{\beta_2}y.\end{aligned}$$

Observe that

$$\begin{aligned}&\int\int_E \left(qa_1^k(x,y)p^{D_{\beta_1}}(x,\beta_2(y)) + qa_2^k(x,y)p^{D_{\beta_2}}(\beta_1(x),y)\right)D_{\beta_1}xD_{\beta_2}y \\ &= \int\int_E \left(\left(qa_1^k(x,y)p(x,\beta_2(y))\right)^{D_{\beta_1}} + \left(qa_2^k(x,y)p(\beta_1(x),y)\right)^{D_{\beta_2}}\right)D_{\beta_1}xD_{\beta_2}y \\ &-\int\int_E \left(\left(qa_1^k(x,y)\right)^{D_{\beta_1}} p(\beta_1(x),\beta_2(y)) + \left(qa_2^k(x,y)\right)^{D_{\beta_2}} p(\beta_1(x),\beta_2(y))\right)D_{\beta_1}xD_{\beta_2}y \\ &= \int\int_\Gamma \left(-qa_2^k(x,y)p(\beta_1(x),y)\right)D_{\beta_1}x + \left(qa_1^k(x,y)p(x,\beta_2(y))\right)D_{\beta_2}y \\ &-\int\int_E \left(\left(qa_1^k(x,y)\right)^{D_{\beta_1}} p(\beta_1(x),\beta_2(y)) + \left(qa_2^k(x,y)\right)^{D_{\beta_2}} p(\beta_1(x),\beta_2(y))\right)D_{\beta_1}xD_{\beta_2}y \\ &= -\int\int_E \left(\left(qa_1^k(x,y)\right)^{D_{\beta_1}} p(\beta_1(x),\beta_2(y)) + \left(qa_2^k(x,y)\right)^{D_{\beta_2}} p(\beta_1(x),\beta_2(y))\right)D_{\beta_1}xD_{\beta_2}y.\end{aligned}$$

Therefore,

$$\iint_E qT^k(p^\beta)D_{\beta_1}xD_{\beta_2}y = \iint_E qa_0^k(x,y)p(\beta_1(x),\beta_2(y))D_{\beta_1}xD_{\beta_2}y$$
$$-\iint_E\left(\left(qa_1^k(x,y)\right)^{D_{\beta_1}} p(\beta_1(x),\beta_2(y))\right.$$
$$\left.+\left(qa_2^k(x,y)\right)^{D_{\beta_2}} p(\beta_1(x),\beta_2(y))\right)D_{\beta_1}xD_{\beta_2}y.$$

This completes the proof. □

We define

$$\widetilde{T}^k(q) = qa_0^k - \left(qa_1^k\right)^{D_{\beta_1}} - \left(qa_2^k\right)^{D_{\beta_2}}, \quad k \in \{1,\ldots,n\}.$$

Theorem 8.7 (The Noether Second Theorem without Transformation of Time) *If the functional $\mathcal{L}$ is invariant under the transformations* (8.9), *then*

$$\sum_{k=1}^n \widetilde{T}^k(\hat{E}_k(L)) = 0 \quad on \quad E^\beta.$$

Proof By Theorem 8.5 and Theorem 8.6, we get

$$\sum_{k=1}^n \iint_E \hat{E}_k(L)T^k(p^\beta)D_{\beta_1}xD_{\beta_2}y = \sum_{k=1}^n \iint_E \widetilde{T}^k\left(\hat{E}_k(L)\right)p^\beta D_{\beta_1}xD_{\beta_2}y$$
$$= 0.$$

Hence and by Lemma 7.1, we obtain

$$\sum_{k=1}^n \widetilde{T}^k\left(\hat{E}_k(L)\right) = 0 \quad on \quad E^\beta.$$

This completes the proof. □

Bibliography

[1] A. Hamza, A-S. Sarhan, E. Shehata and K. Aldwoah. A general quantum calculus, Advances in Difference Equations, (2015) 2015:182.

[2] A. Hamza, A-S. Sarhan and E. Shehata. Exponential, trigonometric and hyperbolic functions associated with a general quantum difference operator, Advances in Dynamical Systems and Applications, Vol. 12, No. 1, pp. 25-37, 2017.

[3] V. Kac and P. Cheung. Quantum Calculus, Springer, 2002.

[4] E. Shehata, N. Faried and R. Zafarani. A general quantum Laplace transform, Advances in Difference Equations, (2020), 2020:613.

Index

β-Liouville Formula, 64
β-Matrix Exponential Function, 62
β-Putzer Algorithm, 79
β-Regressive Matrix, 54
$\mathscr{R}_\beta$, 54
$\mathscr{R}_\beta(I, \mathbb{R}^{n\times n})$, 54
$\mathscr{R}_\beta(I)$, 54

Admissible Functions, 226
Admissible Pair, 142
Admissible Variation, 188, 214, 226
Associated Solution, 96, 131

Basis, 114, 130
Boundary, 21
broken line, 33

component of a set, 33
Conjoined Solution, 96, 130
connected set, 33
continuous function, 19
Continuous Functional, 92
Curve
  Closed, 27
curve, 26, 32
  final point, 26
  initial point, 26
  Jordan, 28
  length, 29
  nonrectifiable, 29
  oriented, 27
  parameter, 27
  parametric equations, 27
  rectifiable, 29
  simple, 28

Darboux β-Integral, 14
Darboux integral
  lower, 14
  upper, 14
Darboux Sum
  Lower, 14
  Upper, 14
Disconjugate Equation, 110
Disconjugate System, 136
domain, 33
Dubois-Reymond Lemma, 231

Equation of Motion, 142
Euler Equation, 142
Euler Necessary Condition, 232
Euler-Lagrange Equation, 197, 232

First Variation, 188, 226
Functional, 91

General Partial Derivative, 5
General Quantum Operator, 1
Green formula, 34

Hamiltonian Differential System, 123
Hamiltonian Matrix, 122, 123
Higher Order Fundamental Lemma of the Calculus of Variations, 240

Invariance under Transformations, 242, 246, 250

Jacobi Condition, 210
Jordan β-Measurable, 21, 22
Jordan β-Measure, 22
  Inner, 22
  Outer, 22
Jordan curve, 28

Lagrangian, 188, 213
Legendre Condition, 204
Leibniz Formula, 12
Line Integral
  First Kind, 30
  Second Kind, 31
line segment
  horizontal, 33
  vertical, 33
Linear Functional, 92
Lipschitz condition, 19
Lipschitz constant, 19

Moore-Penrose Generalized Inverse Matrix, 136

Noether Second Theorem, 254
Noether Second Theorem with Transformations of Time, 246
Normalized Bases, 131
Normalized Solutions, 96

Partial Derivative, 5
  Higher Order, 11
  Second Order, 11
Partition
  Inner, 23
partition of interval, 13
Partition of Rectangle, 13
Picone Identity, 100, 157
polygonal path, 33
Positive Definite Functional, 106
positively oriented fence, 33
  set of type ω, 34
Principal Solution, 95, 131
Proper Weak Local Minimum, 188, 226

Quadratic Functional, 105

Refinement of β-Partition, 15
Riccati Equation, 97
Riccati Operator, 136
Riemann β-Integrable, 23
Riemann β-integral, 17
Riemann Multiple β-Integral, 23

Second Variation, 188, 226
set of type ω, 33
Special Normalized Bases, 131
Sufficient Condition for Positive Definiteness, 173
Symplectic Differential System, 120
Symplectic Matrix, 117, 120

Weak Local Minimum, 188, 214, 226

Printed in the United States
by Baker & Taylor Publisher Services